Digital Broadcasting

Digital Broadcasting
Policy and Practice in the Americas, Europe and Japan

Edited by

Martin Cave

Director, Centre for Management under Regulation, Warwick Business School, University of Warwick, UK

Kiyoshi Nakamura

Professor of Media Industrial Organisation, School of International Liberal Studies, Waseda University, Japan

Edward Elgar

Cheltenham, UK • Northampton, MA, USA

Published by
Edward Elgar Publishing Limited
Glensanda House
Montpellier Parade
Cheltenham
Glos GL50 1UA
UK

Edward Elgar Publishing, Inc.
136 West Street
Suite 202
Northampton
Massachusetts 01060
USA

A catalogue record for this book
is available from the British Library

Library of Congress Cataloguing in Publication Data

Digital broadcasting : policy and practice in the Americas, Europe and Japan / edited by Martin Cave and Kiyoshi Nakamura.
    p. cm.
  Includes bibliographical references and index.
    1. Television broadcasting—Cross-cultural studies. 2. Television broadcasting policy—Cross-cultural studies. 3. Digital television—Cross-cultural studies. I. Cave, Martin. II. Nakamura, Kiyoshi, 1946–
  HE8689.4.D54 2006
  384.55—dc22

2005032812

ISBN-13: 978 1 84542 371 1
ISBN-10: 1 84542 371 2

Printed and bound in Great Britain by MPG Books Ltd, Bodmin, Cornwall

Contents

PART III DIGITAL BROADCASTING AND PLATFORM COMPETITION

Contributors

S. Asai: Associate Professor at Otsuma Women's University, Japan

M. Cave: Professor at Warwick University, UK

C. Cowie: Birkbeck College, London, UK

A. Del Monte: Professor at University of Naples, Italy

L. Di Mauro: Sky TV, Italy

K. Domon: Professor at Waseda University, Japan

G. Fontaine: IDATE, Montpellier, France

H. Galperin: Professor at Universidad de Sans Andrés (Argentina) and Associate Professor at the Annenberg School for Communication at the University of Southern California, USA

K. Hayashi: Vice-President and Professor at the Institute of Information Security, Japan

E. Joo: Professor at University of Kitakyushu, Japan

S. Kapur: Lecturer, Birkbeck College, London, UK

E. Kwerel: Senior Economic Adviser at FCC, USA

J. Levy: Senior Economic Adviser at FCC, USA

H. Mitomo: Professor at Waseda University, Japan

K. Nakamura: Professor at Waseda University, Japan

H. Oniki: Professor at Osaka-Gakuin University, Japan

G. Pogorel: Professor at Ecole Nationale Superieure des Telecommunications, France

N.Tajiri: Associate Professor at Waseda University, Japan

Y. Ueda: Senior Manager, NEC, Japan

Foreword

This volume originated in a conference at Waseda University in Tokyo, chaired by one of us (Nakamura) which heard presentations on the development of digital television from authors from Asia, Europe and the United States. We have since extended the material by adding additional perspectives from a greater number of European countries and from Latin America. A number of thematic issues, particularly relating to platform competition and digital rights, have also been addressed. We believe the volume provides an account of the state of play in a number of major countries, forgotten with analyses of where digital broadcasting is likely to go, and to take the communications sector more widely. We are grateful to Gill Allen for help in preparing the manuscript and to Alice Cave for editorial assistance.

Martin Cave
Kiyoshi Nakamura

Acknowledgements

The research was partly supported by Waseda University Grant for Special Research Projects (2003C-015), Grants-in-Aid for Scientific Research for the Promotion of Science ((B) 174020222) and The Telecommunications Advancement Foundation Research Fund Year 2005.

1. Digital television: an introduction

Martin Cave and Kiyoshi Nakamura

The purpose of this chapter is to set in context the contributions which follow describing the development of digital television in a range of American, Asian and European countries, and also discussing certain general issues in the development of broadcasting in the digital era, including the provision of spectrum, protection of property rights and containment of market power.

The first section outlines the development of digital television broadcasting globally. The second section describes the value chain in digital broadcasting, identifies where market power might be exercised, and discusses possible remedies. The third section discusses some of the issues associated with providing spectrum for digital broadcasting, and the implications of digital switch-over. The final section provides a brief review of the chapters which follow.

WHAT IS DIGITAL TELEVISION AND WHO GETS IT AND HOW?

Hernan Galperin's book on digital television in the UK and the USA eloquently explains in its first paragraph what digital television does:

> the transition from a world of spectrum scarcity, dumb terminals, and one-way services, to a world of on-demand programming, intelligent terminals, and abundant channels – namely, a transition from analogue to digital TV. Heralded as the most important innovation in the history of the industry, digital TV involves the reconfiguration of a sector that, beyond its economic significance, is central to the mechanisms of democratic politics and the evolution of popular culture. This is certainly not the first time that the television industry faces reorganization on a massive scale. But for the most part past technological innovations have spurred evolutionary, not revolutionary change. An old black-and-white TV set would probably be able to pick up several colour TV signals. Analogue cable branded packages of programming called channels. The transition to digital TV is different. It requires a complete retooling of the existing video production and distribution infrastructure, from studio cameras

to transmission towers. It requires new mechanisms to compensate content creators and is distributed in a world where conventional ads can be skipped and perfect copies made and distributed with the click of a button. And it requires new tools for viewers to navigate the maze of programming and new services available, much like Internet browsers help us find our way through the World Wide Web (Galperin 2004, p. 3).

The world may be going down this road, but it is doing so in different proportions and at different speeds in different countries. What we observe is the progressive addition of new services which offer viewers first more channels, made possible by the much greater compression attainable under digital rather than analogue technology;[1] then some interactivity, which can involve an enhanced electronic programme guide (EPG), for example with a search engine, content on demand, opportunities to modify a programme via display of additional information or choice of camera angle, interactive advertising, and use of the television to play games or place bets (Jensen 2005); then, in the more distant future, a more comprehensive change in the economic and social relations linking consumers, content providers, and all the intermediate steps in the processes linking providers and consumers of content.

We will begin with platforms, because these are the most visible aspect of the new digital production processes. Initially we had analogue terrestrial transmission, then (in some countries) low capacity analogue cable, then analogue satellite, offering many channels over a significant 'footprint' served by a transmitter in geo-stationary orbit. Compared with analogue terrestrial, cable and satellite broadcasting was already beginning to chip away at the limitation on access to content supported by barriers to entry arising from limited spectrum availability – or the reluctance to release what there was or the part of governments which were often in thrall to politically influential incumbents.

Transition to digital platforms both expands the capacity of each and breaks down divisions among different types of communication, such as television, voice communication or access to the internet (a process often known as convergence). Table 1.1, adapted from Chapter 6, exposes the range of possibilities – either available or in prospect.

The take up of digital TV is variable throughout the world. Table 1.2 shows take up rates in 2005 in a range of countries. The UK had reached the highest level – nearly 70 per cent. Other countries, such as the Netherlands with nearly 100 per cent cable penetration, had barely made a start. The take up rate of DTT in the United States remained very low – see Kwerel and Levy (Chapter 2) below – although satellite services were completely digital, cable increasingly so and DTT was widely avail-

Table 1.1 Characteristics of digital platforms

Platform	Services	Approximate capacity to home	Interactivity	Strengths	Weaknesses
Digital cable	TV, radio, PPV, interactive services	Equivalent to about 800 TV channels	Good scope – integrated return path	Large bandwidth Integrated return path	Limited return path capacity Fragmented networks add to costs
Digital terrestrial transmission	TV, radio, interactive services	4–6 times analogue capacity (say 40–80 channels)	Limited scope and no return path	Large bandwidth Mobile and portable indoor use (theoretically) Local differentiation possible	No integrated return path Expensive way to achieve universal coverage
Digital satellite	TV, radio, PPV, interactive services	Equivalent to 600–1000 TV channels	Good scope but lacks integrated return path	Large bandwidth National coverage from one satellite Fast roll-out of innovations is possible, for example, personal video recorders and high definition TV	No indoor reception No integrated return path Long transit time doesn't suit all interactive services
DSL based on telephone network	TV, radio, VOD, interactive services, Internet	Unlimited (video on demand)	Excellent – integrated high-speed return path	Large bandwidth Near universal coverage	Bandwidth available drops with distance from the exchange

Table 1.1 (continued)

Platform	Services	Approximate capacity to home	Interactivity	Strengths	Weaknesses
Powerline, based on electricity network	Internet	(as for DSL)	Excellent – integrated high-speed return path	Ubiquitous access extends throughout houses	Not yet rolled out Interference problems still to be resolved
Wide area wireless broadband	Internet	Up to 1 Mbps at present	Good	Extends the reach of fixed broadband platforms	Coverage varies within home
Mobile broadcasting DVB-H	TV, radio, interactive services	30–80 channels per multiplex	Good – uses 2G/3G return path	Suited to mobile TV and other multimedia services Backed by major handset vendors	Not mature yet Needs spectrum opportunities
3G mobile	Voice, Messaging, Audiovisual stream/ download, interactive services,	No live TV at present Typically ~384 kbps	Good	Good for Internet access Fast roll-out within covered areas	Coverage variability

Note: DSL = digital subscriber line; PPV = pay per view; VOD = video on demand.

Table 1.2   Digital TV penetration rates (per cent) end 2005

	Total	Cable	Satellite	DTT	IPTV*
UK	68.9	10.5	32.0	25.2	0.2
France	34.7	4.3	21.6	6.9	1.9
Germany	28.9	6.7	17.8	4.2	0.1
Italy	36.0	0.0	20.2	14.9	0.9
Netherlands	11.4	5.3	3.1	2.3	0.6
Poland	17.9	0.4	17.5	0.0	0.0
Spain	27.6	5.6	15.4	5.2	1.5
Sweden	44.5	9.6	20.6	13.3	1.0
USA	50.3	25.3	24.2	0.5	0.3
Japan	59.1	7.2	33.1	17.9	0.9

Note: *Delivered by DSL or equivalent technology.

Source: Screen Digest.

able. In Japan, analogue terrestrial is firmly targeted for switch-off in 2011.

THE DIGITAL VALUE CHAIN AND MARKET POWER

As shown above, digital television can be delivered on many platforms, although at present cable, satellite, and to a lesser extent DTT, predominate. Moreover, the platform is a single element in a complex value chain which goes from the initial creation of broadcast content by actors, sportspeople, cinematographers, and so on and ends up with the reception equipment in a home or (in the future) on a mobile terminal. Some firms provide nearly all the services involved in the value chain; others participate in only one (for example, programme making or satellite distribution). The outcome for viewers in terms of choice, price and quality depends upon what happens in the chain as a whole.

From the consumer's viewpoint, a key determinant of outcomes is the exercise of market power. The owner of a bottleneck at any point in the value chain will try to appropriate the monopoly rents available throughout.

Table 1.3   Digital value chain

	Content creation	Programme making, filming sports etc
	Wholesaling of programmes	Complete channel or pay per view
	Retailing	Often as 'bouquets' of channels
Platform	Transmission	Cable, satellite, DTT, ADSL
	Conditional access, and so on	Not relevant for 'closed channels' such as cable
	Customer provides equipment	Including STB's – often subsidised.

For this reason competition authorities and regulations have taken a keen interest in the exercise of market power in broadcasting.

We illustrate this in terms of a simplified value chain shown in Table 1.3 (see Cave 1997). Under advertiser-supported broadcasting, instead of programmes being retailed to audiences, the audiences themselves are sold to advertisers. This activity replaces revenue capture via conditional access. Traditional analogue terrestrial broadcasting in Europe, undertaken by monolithic firms, often in public ownership, shows a high degree of vertical integration of these activities.

Throughout the value chain, competition takes place within the framework of standards adopted *de jure* or *de facto* in particular regions. An example is provided by the various versions of the digital video broadcasting (DVB) standard adopted in Europe. There has been much debate about the appropriate role of standards and the consequence of the existing multiplicity of standards across the world (Galperin 2004, Levy 1999). Most of the chapters which follow take existing standardisation arrangements as a given in their discussion, as that phase of activity has passed. But the past (and future) role of standards should not be underestimated.

A more realistic example, loosely based on the state of pay TV in Italy, is shown in Figure 1.1 (see also Chapters 5 and 11 by Del Monte and Di Mauro). This involves separation of ownership (illustrated by dotted lines) and interchange among stages.

In this representation:

(a) content providers sell exclusively or non-exclusively to wholesalers;
(b) the latter make their material available to multiple retailers;
(c) one platform buys all its content at wholesale;

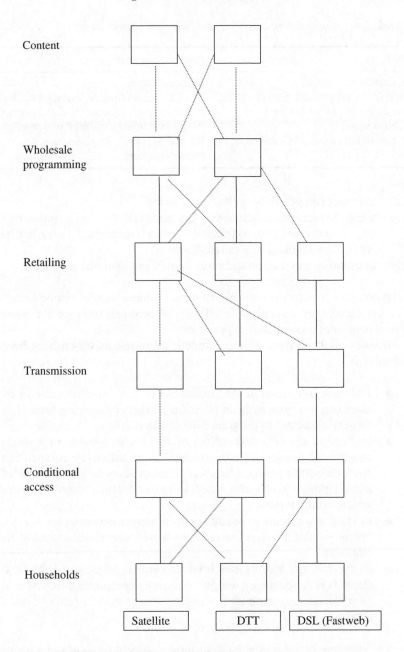

Content

Wholesale
programming

Retailing

Transmission

Conditional
access

Households

Satellite   DTT   DSL (Fastweb)

Notes: Dotted lines represent transaction boundaries.

Figure 1.1   Competition in the value chain: an example

Table 1.4   Broadcasting activities and likely problems

Activity	Competition problem
Content	Monopolisation
Wholesale programme market	Bundling, price squeeze, excessive pricing
Retailing of pay programmes	Bundling, price squeeze, excessive pricing
Transmission	Abuse of dominance, denial of access
Conditional access, EPG and so on	Refusal to supply, discrimination, excessive pricing

(d)   a retailer can get on to multiple platforms;

(e)   using the same or a different delivery mode (satellite, DTT, and so on), any two activities, such as transmission and conditional access, can be vertically integrated or separated; and

(f)   households can take programmes from more than one platform.

This complex structure generates an almost limitless array of opportunities for the exercise of market power. Table 1.4 contains some of the more prominent ones encountered in practice.

By way of illustration of the competition problems which have been identified:

- The issue has arisen in the United States as to whether cable operators can foreclose entry in platform markets by denying their platform rivals access to programming (Hazlett 2005).
- In Europe, the collective selling of live football rights to a single programme wholesaler has attracted intervention by competition authorities; the practice has been impugned as a joint exercise of market power which also leads to reduced competition in downstream retail markets.
- In Italy, it was made a condition for a merger between two pay-TV operators that the combined entity would give rivals access to its platform.
- In the UK the leading pay broadcaster was investigated for (and cleared of) conducting a margin squeeze by preventing purchasers of its wholesale programmes from being able to make a profit in retail markets.

The chapters which follow pick up these competition issues in various ways. They also address questions of how legitimate protection of intellectual property can be achieved. In this context a particularly interesting development is the growth of digital rights management (DRM). Copyright on content can be justified by the need to provide incentives for its creation (Scotchmer 2004). In a digital world, making perfect copies is easy and cheap. Content owners thus have a strong incentive to protect their intellectual property and have a range of new technical options available. As Cowie and Kapur show in Chapter 9, some of these supersede functions now carried out by platform providers, thus altering the balance of market power in the value chain. In Japan the issue of improving a defective copyright regime has become increasingly important, as Chapters 10 and 12 by Domon and Joo and Oniki show.

SPECTRUM POLICY AND THE DIGITAL SWITCHOVER

Public policy in the field of spectrum allocation also exercises a powerful influence on digital TV. Traditionally, governments used their power to assign spectrum as an auxiliary instrument for controlling the number and identity of broadcasters. Traditional spectrum management techniques suited this purpose very well.

These techniques are known as 'command and control' and have operated in essentially the same way since the first global convention for co-ordinating spectrum use in 1906. Under the system, spectrum blocks are allocated, through international agreement, to broadly defined services. National regulatory authorities then assign licences for use of specific frequencies within these allocations within their jurisdictions (Cave 2002, 55).

This regulatory task involved an inherently complex balancing act in a range of dimensions, in each of which there are many conflicting considerations. Key factors include:

- *Interference* Transmissions interfere unless sufficiently separated in terms of frequency, geography and time. Regulators must strike a balance between reducing the extent of harmful interference, through careful planning, and enabling potentially valuable new services to enter the market.
- *International co-ordination* The effective use of radio spectrum in one country will typically require careful co-ordination with neighbouring countries, to mitigate the extent of harmful interference.
- *Investment in equipment* Most radio equipment can operate over only a limited range of frequencies, and so relies on predictable access over time to defined frequency bands. Stability in spectrum assignments to encourage investment in equipment can slow the pace of

spectrum re-use. Increasingly, technical specifications are determined internationally to reap economies of scale in production. National regulators need to balance stability and international harmonisation with responsiveness to new technologies.

The problems of co-ordinating broadcasting spectrum are particularly severe, since broadcasting is a 'one-to-many' communications technology which is efficiently done over a large area. This inevitably creates the risk of interference with broadcasters in neighbouring areas or countries. This problem was vividly exposed in the United States in the 1920s when a Court ruling denied the Government the power to control access to spectrum. The resulting free-for-all, in which radio stations progressively turned up their power to resist interference from others led to a 'Tower of Babel' and eventually to the Radio Act 1927 which gave the Secretary of Commerce power to authorise and control access to spectrum.

The resulting problems are resolved in the age of television broadcasting by international agreements, which set out in great detail which transmissions at what power are permissible from which specified sites. Thus analogue terrestrial television broadcasting in Europe is governed by agreements reached in Stockholm in 1961. A European plan for digital broadcasting is currently being developed for approval at a Regional Radio Conference in 2006.

Subject to these constraints, each national spectrum authority assigns frequencies to particular broadcasters. For example in the United Kingdom, analogue and digital terrestrial TV transmissions use 368 MHz of spectrum within the band 470–854 MHz (Cave 2002, 161). The spectrum is split into 46 frequency channels, each composed of 8 MHz of spectrum. The following bands are used:

- 470 to 590 MHz (channels 21 to 35);
- 598 to 606 MHz (channel 37);
- 614 to 854 MHz (channels 39 to 68).

(To complicate matters, channel 36 is allocated to radar for historical reasons.) Each channel can be used to broadcast either one analogue TV service, or one digital multiplex – carrying six or more separate TV services – from a given transmission site. There is a maximum of 11 channels used at a transmission site (five for analogue TV channels, and six for DTT multiplexes). At such sites there are still seemingly 35 frequencies (46 minus 11) lying idle. These empty frequencies are interleaved with the frequencies used for the analogue and digital services. Some of the empty interleaved frequencies' channels cannot be used because they would cause interference with the channels which are used or with adjacent transmitters; some, however, could be made available to broadcasters or other users.

Satellite broadcasting also requires spectrum for two purposes – to uplift signals to the satellites and to broadcast signals direct to home (DTH). As signal strength from the medium-powered satellites currently in use is fairly low, frequencies must be cleared of alternative services to allow signals to 'get through'. A further feature of satellite broadcasting is that, because transponders have a multinational footprint, and because uplift and reception can be in different countries, the spectrum authority in the country where the signal is received may have no jurisdiction over the provider of transmission services.

While command and control has been used almost universally for managing broadcasting, and most other frequencies, the attention has turned in several countries to the alternative of using market mechanisms to allocate and assign spectrum. A start has been made in this process by the use of auctions to assign licences, especially for third generation mobile services, but the market agenda extends to 'secondary trading' – that is exchange of ownership of spectrum or spectrum licences which have already been issued, accompanied by the opportunity for the existing or new licensee to change the use of the spectrum – often known as liberalisation, subject, of course to international obligations. The US and UK spectrum management agencies have supported and to some degree introduced secondary trading and liberalisation (FCC 2002, Ofcom 2005).[2] The European Commission is expected shortly to endorse secondary trading.

Under a liberalised regime, spectrum can be used interchangeably, for example, for mobile communications, traditional terrestrial broadcasting, mobile broadcasting and a range of other possibilities. The chosen use would depend, as in other markets on the varying willingness to pay of potential licensees, which would itself depend, where end users are buying the services, on their willingness to pay for those services. As noted below the switchover to digital terrestrial will confront spectrum authorities with a choice between command and control and market methods of assigning released spectrum.

One consequence of the switchover to digital terrestrial is that, for a transitional period, both analogue and digital platforms have to be used at once. The length of the period is under government control, but the turnover of customer premises equipment – televisions, VCRs, and so on – of which there may be three or more per household – is a slow process, and provision may have to be made to encourage the acquisition of digital set top boxes by slow adopters. (Such policies are analysed in Adda and Ottaviani (2005).) In Italy, which has a switch-over target date (almost certainly unrealistic) of 2006, the Government offers set top box subsidies of 40 euros per household. Cave's chapter describes how the issue is being faced in the UK. Clearly, government to pre-announce a subsidy if it will choke off investments by individual households.

None the less, the possible 'spectrum dividend' associated with analogue switch-off has encouraged most governments of richer countries to seek a digital switchover of terrestrial transmission, which both brings the other advantages of digital television noted above and releases valuable spectrum which can be assigned to other users, either by command and control or market methods.

The European Commission has proposed a target date of 2012 for the 20 member states of the EU, setting 2015 as the latest date for members. Table 1.5 shows the state of the process in each country.

Digital switchover in the USA is now scheduled for 2009 under recently passed legislation. In Japan a plan for digital terrestrial broadcasting has been implemented since 2003, starting with the three main metropolitan areas of Tokyo, Osaka and Nagoya. It is planned that analogue broadcasting will cease on 24 July 2011.

THE CHAPTERS WHICH FOLLOW

The interplay of forces which determine a country's broadcasting system tends to be much more idiosyncratic than those which govern, for example, its telecommunications sector.[3] The chapters which follow therefore take up a variety of themes. First, six chapters provide a number of country or area studies; further country studies of digital terrestrial television in Europe can be found in Brown and Picard (2005). Then a group of three chapters discusses rights issues. Finally, four chapters analyses various aspects of platform competition in digital broadcasting.

Chapter 2:   The DTV transition in the US (Kwerel and Levy)

This chapter discusses the transition to digital television (DTV) in the US on three platforms: terrestrial broadcasting, cable and satellite. DTV provides more and better services for consumers on all three of these platforms, and promotes inter-modal competition, as without the transition of terrestrial broadcast television from analogue to digital it is questionable whether terrestrial broadcast television could survive as a viable competitor to cable and satellite. The transition of broadcast TV to digital also has the additional benefit of freeing up spectrum for other potentially more valuable uses.

Table 1.5   European digital switchover timetables

Country	Target Date	Status	Other details
Austria	2010	Provisional	
Belgium	2012	Provisional	SO planned to start in 2010
Cyprus	no date set		
Czech Republic	2017	Provisional	
Denmark	2011	Provisional	No decision yet
Estonia	no date set		
Finland	2007	Provisional	August 2007 – all of country
France	no date set		No decision yet
Germany	2010	Provisional	Berlin switched off 2003.
Greece	2010	Provisional	
Hungary	2012	Provisional	
Ireland	no date	Provisional	No decision yet
Italy	2006	Committed	
Latvia	2006	Committed	
Lithuania	2012	Provisional	SO starting in 2012
Luxembourg	no date set		No decision yet
Netherlands	no date set		No decision yet
Malta	no date set		
Poland	no date set		
Portugal	2010	Provisional	No decision yet
Slovakia	2015	Committed	2012 target
Slovenia	2015	Committed	2012 target
Spain	2010	Provisional	May start earlier by region
Sweden	2008	Committed	February 2008 – proposed final date.
United Kingdom	2008–2012	Committed	

Source: European Union, May 2005.

In assessing the current status of the US DTV transition, a number of indicators can be considered. One of these is the number of stations actually transmitting in digital format. The authors find that terrestrial DTV signals are widely available throughout the US, and that digital programming is also widely available on cable. Note that satellite television is a completely digital service. Another factor to consider is the availability of digital content via the various distribution platforms. All the major US commercial broadcast television networks are currently producing significant amount of HDTV content, and there are roughly 17 non-broadcast HDTV services distributed by cable and satellite, representing a variety of genres. The take-up of DTV equipment is also an indicator, and sales figures suggest that approximately 3.3 per cent of US households had the capability to receive terrestrial DTV signals off-air at the end of 2004.

Digital rights management (DRM) policy has a strong impact on DTV diffusion, as consumer demand for DTV equipment depends heavily on the content available and the willingness of content providers to make content available depends on their ability to retain control over distribution and, especially, to prevent or limit unauthorised copying. Unlike analogue material, digitally formatted data may be copied repeatedly with little or no quality degradation, diminishing the return to the creators of the content and thus diminishing the amount of content produced. This reduces the incentive for consumers to invest in DTV equipment. Policymakers must balance incentives to produce new content against the potentially substantial benefits to consumers of copying and sharing.

A target date of the end of 2006 had been set for switching off all analogue television in the US. However, the legislation permitted stations to continue analogue broadcasting in markets where penetration of DTV reception equipment is less that 85 per cent. In 2006, legislation was passed which set a hard date of 17 February 2009 for switchoff regardless of penetration rates achieved. Funding for convertor box subsidies was approved up to $1 bn, with a provision for an increase to $1.5 bn. The auction of spectrum recovered for commercial use must start by January 2008.

Switching to digital broadcasting would free up spectrum that could be valuably used elsewhere, but the cost of the loss of analogue service to viewers some or all of whose sets rely on analogue signals must be taken into account. However, this cost is much less than the value of the spectrum made available for other uses. Therefore, incentives should be provided to speed up the switchover.

Chapter 3: Digital broadcasting in the developing world: a Latin American perspective (Galperin)

This chapter examines the issues associated with the switch to DTV in the developing world, specifically in Latin America. While the transition to DTV has been underway for some time in most of the developed world, Latin American nations have only recently begun to set the terms for the transformation, which implies a complete retooling of the existing production and distribution systems, as well as new mechanisms to compensate creators and distributors for the increased ease of copying, which makes new demands on the existing legal apparatus.

Some of the advantages of DTV are particularly relevant to the developing world. First, it may help fuel the growth of wireless communication infrastructure, critical in nations where no legacy of wired networks exists. Second, it can offer a unique gateway into the information society for those who are currently excluded: while few households in Latin America have computers or internet connections, most have televisions.

However, there are also challenges involved in the transition to DTV, as the experiences of developed nations have demonstrated. Galperin identifies two distinct strategies that have been adopted to overcome these obstacles: the US approach in which the existing broadcasting arrangements are extended into the digital era and the approach exemplified by the UK, where there has been a reallocation of spectrum rights and greater government involvement in managing the transition and addressing potential co-ordination problems. Countries in Latin America must consider both local and international factors in deciding which approach to adopt: the proximity of some of the nations to the US has proved an important deciding factor. Galperin goes on to focus more closely on the markets of Argentina, Brazil and Mexico, three key Latin American markets.

Chapter 4: DTT and digital convergence: a European policy perspective (Fontaine and Pogorel)

Television policy in the EU has been designed to conform to overall aims of European integration. This requires co-ordination of national legislation over an area where the specifics of the broadcasting markets of individual countries often differ significantly. In larger countries (with the exception of Germany), terrestrial transmission tends to be the largest platform. Conversely, it is marginal in almost all of the smaller European nations, such as Benelux, Switzerland and Scandinavia, which have opted for cable or satellite as it has not been feasible to develop a broad national offer on terrestrial broadcasting. Terrestrial migration to digital could alter the situation by stimulating a resurgence of interest in terrestrial broadcasting as the range of services offered expands, making it a viable alternative to cable or satellite, and possibly increasing usage, especially in countries where terrestrial broadcasting is currently secondary. However, in many of the larger European countries digital terrestrial television will not cover as much of the population as analogue (often as little as 85 per cent) so a greater number of people will have to resort to an alternative platform.

Chapter 5: The development of digital broadcasting in Italy (Del Monte)

There are a number of common justifications for regulating the broadcasting market as both economic and social. However, the policy adopted with regard to regulation differs in different countries and this has resulted in different market structures for the analogue TV sector. These structural features constrain the transition from analogue to digital to the extent that the transformation often follows a pattern determined by the pre-existing industry structure. Therefore the amount and type of regulation needed to maximise the benefits of the introduction of digital television varies from country to country.

In theory, the relaxation of spectrum scarcity should decrease barriers to entry and increase competition in the television market, and the higher number of channels will increase product differentiation. However, it is unlikely that market forces alone will bring about these changes. For example, if the allocation of licences is not designed to favour new entrants, the incumbent operators, who benefit from key assets such as distribution infrastructure and ownership of content, will preserve and ultimately strengthen their market position as the transition to digital takes place.

The pay-TV market is much less developed in Italy than in other European countries due to features of the regulatory environment from the 1970s onwards, so DTT faces little effective competition from other platforms. This means that the Italian media laws alone are incapable of increasing competition in the digital service markets, as are other rules of conduct which might be sufficient in other national broadcasting markets. In Italy structural regulation will be necessary to achieve the goal of a genuinely competitive market.

Chapter 6: The development of digital television in the UK (Cave)

The UK has a 2005 penetration level of digital TV of 60 per cent, and plans to switch off analogue terrestrial transmission on a regional basis from 2008–2012, and immediately to achieve the same level of near-universal coverage by DTT.

Throughout the 1990s, multi-channel television, delivered by cable and satellite platforms began to take off, and in 1998 the Government licensed DTT, broadcasting to about 70 per cent of households a mixture of non-pay and pay services. However the pay-TV licence did not prosper and in 2002 a multi-channel largely free-to-air service was launched (Freeview) which quickly acquired 5 million customers. In the coming years, cable, satellite and DTT are likely to compete with DSL-based services.

A further boost to DTT is provided by digital switchover, as the Government has required that (almost) all households are provided by a DTT service. The switchover will take place progressively over a four-year period beginning in 2008.

The use of the freed analogue spectrum will be decided within the framework of the UK's spectrum management policies, which over the past decade have sought increasingly to impose market-type disciplines on all spectrum users including broadcasters, and now propose to rely primarily on auctioning and secondary trading of spectrum (allowing change of use) to make spectrum assignments. As a result, released analogue spectrum may be deployed in a variety of users, including DTT, mobile television (known as DVB-H) or mobile communications.

Chapter 7: A perspective on digital broadcasting in Japan (Nakamura and Tajiri)

Japan is undergoing the same transition to digital television as is occurring in other countries. Three principal platforms are in play: DTT, which was inaugurated in 2003 and should supersede analogue terrestrial transmission in 2011; satellite broadcasting, using both 'broadcasting' and communications satellites, which are switching to digital; and cable, where digitalisation is still in its infancy. In addition, fibre to the home installed by telecommunications companies provides a fourth platform.

The chapter focuses upon relations between content providers and broadcasters. The former are subject to 'natural oligopoly' tendencies as a result of the superiority of particular forms of content – vertical quality differentiation. However certain content providers are subject to potential abuses from more powerful incumbent and protected broadcasters.

This latter problem finds reflection in copyright issues, which tend to be resolved in favour of broadcasters, as a result of inadequate procedures, especially for secondary rights.

A further problem exists with spectrum policy, where current assignment and charging regimes fail to either promote economic efficiency or to achieve equity between under-paying broadcasters and over-paying mobile operators.

However the authors conclude that the developing features of the communications sector – notably convergence – provide an encouraging context for the development of a new competition regime based upon competition with existing new markets and competition for innovative ones.

Chapter 8: Legal and economic issues of digital terrestrial television (DTTV) from an industrial perspective (Hayashi)

This chapter discusses the issues associated with DTTV from an industrial perspective by taking the Japanese experience of DTTV as an example. The shift to DTTV can be viewed in one of two ways. Under the substitution hypothesis DTTV is nothing more or less than broadcasting based on digital technology rather than analogue. This is the view of DTTV subscribed to by most broadcasters and policymakers. Hayashi offers the paradigm shift hypothesis as the view of DTTV diametrically opposed to the substitution hypothesis, whereby DTTV is in fact radically different to conventional broadcasting, necessitating a rethink of the entire broadcasting industry. Namely, content and conduit, which have traditionally been vertically integrated, should now be treated separately. In addition, transmission using Internet Protocol should be implemented to respond to the diversity in conduits and receivers.

The experience of the internet revolution and the convergence of communications and computing that it entails suggests that although dependence on the substitution hypothesis as a business plan may provide a solid start, it may sometimes be overtaken by the paradigm shift hypothesis, as was the case with the choice of routing technology. However, the paradigm shift hypothesis essentially provides policymakers with unlimited scenarios, while the substitution hypothesis automatically provides limits and thus lends itself better to analysis of the issues involved in the transition to DTTV. Hayashi goes on to discuss some of these issues, but then states that the ensuing analysis highlights the inadequacy of the substitution hypothesis, again suggesting that we should look at the transition to DTTV in terms of a paradigm shift rather than in terms of evolutionary change.

Chapter 9:   The management of digital rights in pay TV (Cowie and Kapur)

At present, broadcasting content is protected by a mixture of legal and technological measures. The transition to digital television poses new challenges for the protection of this content as piracy becomes easier, prompting both changes to copyright law and new technological solutions. Digital Rights Management (DRM) is one such technological response, one which the authors believe will be increasingly important for protecting digital content in the pay TV supply chain. This chapter examines the potential impact of DRM on market power in digital pay TV.

DRM is a suite of technology employing a number of discrete tools which combine to allow the content owners to monitor and control access to material from the point of creation to the end of its life. Preventing unauthorised duplication and transfer enables the owners to appropriate the full value from content, providing greater incentives to support the new digital platform, and thus welfare gains arise in the form of new products.

DRM also reduces the transactions costs associated with the transfer of digital media. This renders new business models viable and leads to a change in the economics of the supply chain at intermediate levels as content owners can threaten to use DRM to distribute content directly to consumers, bypassing pay TV platforms. This has implications for regulation and competition policy.

DRM also improves the ability to charge discriminatory prices. This could potentially be welfare improving, but price discrimination often causes dissatisfaction in consumers, sometimes translating into calls for regulatory intervention.

Chapter 10: Copy control of digital broadcasting content: an economic perspective (Domon and Joo)

This chapter adopts an approach to controls over copying digital broadcasting material based on the economic literature. The latter has analysed in particular the pros and cons of allowing copying of printed matter, reaching ambiguous conclusions both theoretically and empirically. A key issue has been the extent of indirect appropriability – the degree to which the content provider can appropriate copiers' willingness to pay via an increase in the demand price of the direct customer.

The same phenomenon applies in respect of pay-TV products. While personal copying benefits both producer and consumer, by increasing consumption, copying by third parties increases competition in the market while creating opportunities for indirect appropriability.

The authors concluded that digital broadcasters will be forced to confront the same issues as has the music industry. Copyright infringements will not disappear, and this will force action to be taken. But those actions must take account of the effects on indirect appropriability.

Chapter 11: Regulation of digital TV in the EU: divine coherence or human inconsistency? (Di Mauro)

Regulation of content and regulation of infrastructure are treated separately within the EU regulatory framework. This article focuses solely on the regulation of infrastructure, specifically on infrastructure needed to convey digital TV broadcasting signals.

The most important aspect of the regulation of DTV infrastructure is the regulation of conditional access systems (CAS). CAS is any system or arrangement that allows a broadcasting service to be consumed only by those who are entitled to consume it, usually through subscription. The introduction of CAS represents a shift in broadcasting from a quasi-public good status to that of a standard good as CAS displays the key feature of excludability.

The basic EU regulation of CAS, dating back to 1995, states that providers must offer their services on terms that are fair, reasonable and non-discriminatory (FRND). This was supplemented in 2003 by the new regulatory framework (NRF), based more explicitly on economic analysis than previous regulation. The chapter goes on to examine the determinants of digital broadcasting regulation in the EU, with reference to the FRND structure, especially the discriminatory aspect that is particularly relevant to CAS. The author concludes that although there are dangers associated with adopting a FRND remedy for digital television services, a reasonable interpretation of it is necessary if the needs of end users are to be put at the centre of the regulatory framework.

Chapter 12: Platforms for the development of digital television broadcasting and the internet (Oniki)

The chapter is concerned both with regimes for generating revenue from content and with competition at various points of the value chain in a world where the potential of digital television broadcasting has been realised so that it competes with the internet.

The author first notes that DTT in Japan is being developed in a highly controlled way, which maintains the position of the key network stations in the analogue world. As the dominant source of revenue from television is terrestrial television, its owners seek to maintain the *status quo*, under which revenue comes from advertising, not pay TV.

However, digital television is a technology which is capable of selling content, and the author describes a number of ways in which this could be done, by creating a series of 'windows', subject to disclosure of marketing plans.

The author then turns to problems associated with bottlenecks in the value chains of digital television and the internet. A 'layer' model is developed in which monopoly of infrastructure plays a key role – the situation in Japan in broadcasting and telecommunication markets providing examples of abuses by vertically integrated companies. A variety of antidotes for such conduct are put forward, including obligations to supply at a competitive price, accounting for ownership separation, and competitive interventions by one or more public corporations.

Chapter 13: Economies of scale, scope and vertical integration in the provision of digital broadcasting in Japan (Mitomo and Ueda)

The impending development of digital terrestrial television, which is being introduced at considerable cost with a view to switching off analogue broadcasting in 2011, focuses attention on the cost function of broadcasting firms. The chapter analyses the evidence for Japan, using a sample of analogue broadcasting stations, and testing for economies of scale, economies of scope and the effects of vertical integration.

Economies of scale are investigated using a translog cost function. The results suggest that a broadcaster operating over a larger geographical area is more efficient and this may provide an argument against the tight merger controls at present operating in the industry.

Vertical integration and economies of scope between programme production and transmission is investigated via estimation of cost functions for both activities and for programme production alone.

The policy implications of these results are then examined. A future reorganisation might involve fewer, larger stations, without the vertical integration which has characterised the industry.

Chapter 14:   Comparative analysis of the market structure of broadcasting and telecommunications in Japan (Asai)

The broadcasting and telecommunications sectors in Japan differ in the degree of vertical integration, level of competition and date of digitisation. The chapter examines these relationships, using a general description of economic activity in terms of vertical levels or 'components' and the interfaces between them. Open interfaces are often governed by standards, which in telecommunications are typically *de jure* standards, while the internet relies on a *de facto* standard, TCP/IP.

Broadcasting in Japan is done on terrestrial, cable and satellite platforms. Terrestrial broadcasting is dominant and highly regulated to achieve vertical integration of programme making and transmission and the attainment of a degree of localism within a network-affiliate structure. Satellite markets exhibit more separation and higher levels of competition. Copyright law is undeveloped, although digitalisation is likely to change this.

Japanese telecommunications have been opened up to competition and diversified firms such as KDDI and Softbank have emerged through mergers. The historic monopolist, NTT, has also been legally separated. Digitalisation was completed in 1997.

The author anticipates that when digitalisation of broadcasting is complete, new competition will emerge at the different levels and the dominance of broadcasters over production companies will be reduced. Terrestrial broadcasting will, however, be subject to little internal competition, and competitive processes are likely to come from outside, from sources such as the internet.

NOTES

1. Digital television permits the broadcasting of approximately four to six times as many channels of current definition, or the same number of channels of high definition television, as can be broadcast on analogue with a given amount of spectrum. It also permits widescreen pictures, more robust technical quality and interactive features. If a household owns an analogue set, it must acquire a set top box to convert the broadcast digital system into analogue format.
2. In this respect, they are following earlier pioneers, such as Australia, Guatemala and New Zealand. Further details of UK and USA policy are described in Chapters 2 and 6 by Kwerel and Levy and Cave.
3. It would perhaps flatter telecommunications to recall here Tolstoy's remark that 'all happy families resemble one another, but every unhappy family is unhappy in its own separate way.'

REFERENCES

Adda, Jerome and Ottaviani, Marco (2005) 'The transition to digital television', *Economic Policy*, **41**, 159–209.

Brown, Allan and Picard, Robert J. (eds) (2005) *Digital Terrestrial Television in Europe*, Mahwah, NJ: Lawrence Erlbaum Associates Publishers.

Cave, Martin (1997) 'Regulating digital television in a converging world', *Telecommunications Policy*, **21**, 575–596.

Cave, Martin (2002) *Review of Radio Spectrum Management*, London: HM Treasury and DTI.

FCC (2002) *Spectrum Policy Taskforce Report*.

Galperin, Hernan (2004) *New Television, Old Politics: the Transition to Digital TV in the United States and Britain*, Oxford: Oxford University Press.

Hazlett, Thomas (2005) 'Cable television' in S. Majumdar, I. Vogelsang and M. Cave (eds), *Handbook of Telecommunications Economics, Vol. 2*, Amsterdam: Elsevier.

Jensen, Jens J. (2005) 'Interactive content, applications or services', in Allan Brown and Robert J. Picard (eds), *Digital Terrestrial Television in Europe*, Mahwah, NJ: Lawrence Erlbaum Associates Publishers, pp 101–132.

Levy, David (1999) *Europe's Digital Revolution: Broadcasting Regulation the EU and the Nation State*, London: Routledge.

Ofcom (2005) *Spectrum Framework Review*.

Scotchmer, Suzanne (2004) *Innovation and Incentives*, Cambridge, MA: MIT Press.

PART I

The development of digital broadcasting

2. The DTV transition in the US

Evan Kwerel and Jonathan Levy[1]

INTRODUCTION AND BACKGROUND

All communications, whether video, voice, data, wired, or wireless, are going digital. This chapter focuses on the transition to digital television on three platforms: terrestrial broadcasting, cable and satellite, with particular attention to terrestrial broadcasting in the United States. Digital television, regardless of the platform, provides more and better services for consumers, including HDTV.

The digital migration of television also promotes inter-modal competition. Without the transition from analog to digital it is questionable whether terrestrial broadcast television could survive as a viable competitor with cable and direct broadcast satellite (DBS).

The digital transition of terrestrial broadcast TV has an additional benefit not associated with the two other platforms. Phasing out analog television broadcasting will facilitate more efficient use of the spectrum. Using digital technology allows for much greater output with less spectrum input, freeing up spectrum for other potentially more valuable uses. Those uses are likely to be next generation mobile, including mobile data and video. If a country goes through the transition to digital broadcast television without freeing up spectrum for other uses, it has lost a major benefit of this transition.

The digital television (DTV)[2] transition in the US will clear 108 MHz of spectrum for other valuable uses. Pursuant to the 1997 Balanced Budget Act (BBA), the FCC reallocated 108 MHz (channels 52–69) of the 402 MHz that had been allocated to terrestrial television broadcasting, leaving 294 MHz for TV. Twenty-four MHz of the 108 MHz was allocated for public safety and 6 MHz, which has already been auctioned, was allocated for commercial guard bands. The rest was allocated for flexible use – 30 MHz in the upper 700 MHz band[3] and 48 MHz in the lower 700 MHz band,[4] 18 MHz of which has already been auctioned. The 700 MHz band is adjacent to the cellular band.[5] Even though this 108 MHz of spectrum has been reallocated it is still largely occupied by broadcasters. In the lower

700 MHz band there are 94 analog TV stations and 168 DTV stations. And in the upper 700 MHz band there are 97 analog TV stations and 20 DTV stations. Completing the DTV transition would make this spectrum available for highly valuable commercial and public safety uses.[6]

In the United States, the government has been heavily involved with managing the transition from analog to digital television. After adopting the ATSC standard for terrestrial digital television ('DTV') transmission in 1996,[7] the FCC set 2006 as a target date for completing the transition, with provision for reviewing this decision every two years.[8] The 1996 Telecommunications Act provided that, if the Commission issued 'additional licenses for advanced television services', they should go to existing licensees or permittees, and the Balanced Budget Act of 1997 set a target deadline of 31 December 2006 for switching off analog television.[9] However, the legislation permits television stations to retain their analog authorization beyond that date in markets where household penetration of DTV reception equipment is less than 85 per cent.[10] Recent legislation requires the end of analog television broadcasting by 17 February 2009 and provides for a digital-to-analog converter box subsidy.[11]

The next section reports on the current status of the US digital television transition, while the following section addresses issues related to completing the transition. The final section of the chapter contains conclusions.

CURRENT STATUS OF THE DTV TRANSITION IN THE US

There are three related metrics by which we can measure the progress of the DTV transition. First, there is the number of stations actually transmitting in digital format. Second, there is the availability of digital content via the various distribution platforms, in particular, terrestrial broadcasting, cable television, and Direct Broadcast Satellite (DBS) service. Third, there is the takeup of digital reception and display equipment. This section discusses them in turn and then addresses the issue of digital rights management, which has a direct impact on content availability and equipment manufacture.

Digital Signal Availability

Digital terrestrial television signals are widely available throughout the United States. In the top 30 television markets, 100 per cent of the affiliates of the top four commercial networks are transmitting in DTV format. Overall, 1525 DTV stations are in operation, comprising 89 per cent of all

DTV allotments. The figure for noncommercial educational stations is 88 per cent, only slightly lower than the overall average.[12]

Digital programming is also widely available on cable. The cable industry estimates that 98.7 per cent of the 109.6 million US television households are 'passed' by cable (that is, it is available to them if they are willing to pay the subscription fee).[13] Ninety million US television households are passed by HDTV service, and some additional households are passed by digital cable that does not offer HDTV service, but precise figures are unavailable. Approximately 25 million households subscribe to a tier of digital cable programming.[14]

Digital cable programming consists of some retransmitted terrestrial broadcast signals and some cable networks. With regard to local broadcast signals, cable industry sources state that, as of 1 January 2005, 504 of the 1481 DTV stations then in operation were being carried by local cable systems.[15] Currently most DTV channels are not subject to cable signal carriage requirements ('must-carry'). Those commercial DTV signals that are being carried by cable are transmitted pursuant to voluntary 'retransmission consent' agreements between the stations and the cable operators. In early 2005, the Association of Public Television Stations and the National Cable and Telecommunications Association announced an agreement that covers or will cover the signals of all 'local must-carry' public television stations in the United States.[16] Moreover, this agreement provides, subject to some limitations, for cable operator retransmission of up to four separate program streams of free non-commercial digital programming from one station per market during the transition and for multiple stations posttransition. Recently the FCC declined to require cable operators to retransmit multiple digital program streams of any terrestrial DTV station.[17]

Direct Broadcast Satellite service is a completely digital service, with each of the two major providers, DirecTv and Echostar, offering around 300 digital channels to subscribers. With regard to local broadcast signals, DBS carriers retransmit local analog signals by converting them to standard definition digital. In general, local digital terrestrial signals are not now retransmitted by DBS carriers. FCC signal carriage rules for digital terrestrial signals on DBS have not yet been finalized.

DTV Content

The consumer decision to acquire DTV equipment depends on the availability of digital content. Consumers presumably consider the full range of content options available – terrestrial DTV, cable, DBS, and other multichannel video program distribution platforms, and pre-recorded media. Currently, all of the major US commercial broadcast television networks are producing significant amounts of HDTV content, particularly in prime time. There are roughly 17 non-broadcast HDTV services distributed by cable and satellite (although not every cable or satellite service necessarily carries every HDTV program service). Among the genres represented are sports (ESPN), movies (HBO, Showtime), nature and science (Discovery), and general entertainment (TNT).

Some broadcast digital television capacity is being used for 'multicasting'. Multicasting refers to transmission of several program streams (generally all SDTV but can include one HDTV and one or more SDTV) in a single 6 MHz DTV channel. As noted above, the FCC has recently reaffirmed its decision that cable signal carriage requirements apply to only one video program stream per channel, even if that channel is multicasting, but the cable television trade association and the organization of noncommercial educational television stations recently concluded an agreement that provides for cable carriage of multiple program streams from public stations.

DTV Equipment

According to the Consumer Electronics Association (CEA), cumulative sales of all DTV equipment in the United States were 17.9 million units, as of the end of 2004. The total is for factory sales to dealers and so is only an approximation of the quantity of equipment in consumers' hands. The total includes three types of equipment – integrated DTV receivers (that is, with a ATSC tuner and high-resolution monitor); standalone 'DTV-ready' high resolution monitors; and standalone DTV tuners, or set-top boxes. The standalone monitors are not able to receive terrestrial DTV signals and are apparently purchased for viewing DVDs, for gaming and, in some cases, for use with digital cable or satellite services. As of the end of 2004, roughly 3.6 million DTV tuner-equipped units of DTV equipment had been sold, so approximately 3.3 per cent of US television households had the capability to receive terrestrial DTV signals off-air.

Until 2004, most DTV units sold have been the standalone high-resolution monitors, but this is changing, with the CEA predicting that almost two-thirds of the 16.5 million units expected to be sold in 2005 will have a DTV tuner. This is due to a variety of factors. First, consumer demand for DTV reception appears to be increasing. Second, FCC regulations that require ATSC tuners in new television receivers are phasing in. Specifically, as of 1 July 2005 all new receivers with screen sizes 36 inches and above, and 50 per cent of new receivers with screen sizes 25 to 35 inches had to have ATSC tuners included. The phase-in continues with the requirement that 100 per cent of new receivers with screen sizes of 25 to 35 inches have ATSC tuners as of 1 March 2006. Effective 1 March 2007, all television receivers and other video devices (for example, videocassette recorders and digital video recorders) must include an ATSC tuner.[18]

The third factor leading to an increase in ATSC tuner-equipped DTV equipment is the Commission's regulations regarding compatibility of digital cable systems and DTV home equipment. Any receiver labeled as 'digital cable ready' is required to include an ATSC tuner in it, and the CEA estimates that three million such receivers will be sold in 2005. Finally, the price of ATSC tuners has been falling.[19]

Digital Rights Management

Consumer demand for DTV equipment depends heavily on the range of content accessible. In turn, content providers' willingness to make digital content available depends on their ability to maintain control of distribution and, in particular, to prevent or limit unauthorized copying and redistribution. For these reasons, digital rights management (DRM) policy has an impact on the rate of DTV diffusion.

Content owners are reluctant to make their content available without some protection against unauthorized copying and retransmission. Unlike analog material, video transmissions in digital format may be copied repeatedly with little or no quality degradation and may be widely and inexpensively redistributed via the Internet. If users are able freely to make and retransmit perfect copies of digital content, the return to the creators of that content will be significantly diminished and the supply of new digital content reduced. And without digital content, consumers don't want to invest in buying digital reception equipment.

How much protection to provide content owners requires a tradeoff. Policymakers have to balance incentives to produce new content with the benefits of making existing content widely available. The benefits to consumers of copying and sharing content can be substantial. In the analog world, viewers value the opportunity to copy video programming and have developed certain expectations regarding their ability to copy for various purposes, including time-shifting and portability. The VCR and now personal digital video recording devices are very important to consumers. If consumers could never make a copy, much of the benefit of the digital revolution could be lost. But if copying is too easy, then the content providers are going to provide less content. The analysis is analogous to determining optimal patent protection. Increasing the length of patents provides greater incentives to innovate while raising prices and suppressing use of inventions. There is no simple formula for balancing these conflicting objectives.

The FCC has dealt with these challenges in two major proceedings. The 'Plug and Play' Order, adopted in September 2003, addresses subscription

MVPD programming. The 'Broadcast Flag' Order, adopted two months later, sought to impose more limited DRM protection for broadcast programming. However, the US Court of Appeals for the DC Circuit recently overturned the Broadcast Flag rules as outside of the Commission's statutory authority.[20]

The 'Plug and Play' Order[21] has its genesis in Section 629 of the Communications Act. This section, added as part of the 1996 Telecommunications Act, directs the FCC to adopt rules that ensure commercial retail availability (from sources unaffiliated with programming distributors, for example, cable operators) of 'navigation devices' (equipment to access and select multichannel video programming) without compromising signal security. The term 'navigation device' includes television receivers, set-top boxes, and certain other devices. In response, the Commission adopted rules requiring multichannel video programming distributors (MVPDs) to make available to subscribers a separate 'security module' that could function with a set-top box or other navigation device that the subscriber acquired at retail independently. In this fashion, the MVPD could maintain control of signal security and the subscriber could have the benefit of competition in the provision of navigation devices.[22]

In order for the independently-produced navigation devices to interoperate with cable operator-supplied security modules, it is necessary for the navigation device manufacturer to have a license to utilize the proprietary technology associated with the security module. Initially, the cable industry owners of that technology wanted to include in the license the requirement that the navigation device recognize and pass along whatever DRM instructions were associated with the content received. Consumer electronics manufacturers of navigation devices were reluctant to 'sign a blank check', that is, to agree to whatever DRM requirements that content providers might choose to include in their distribution agreements with cable operators. The manufacturers feared 'excessive' limits on home copying.

The DRM rules that the FCC adopted in the 'Plug and Play' proceeding are based on a proposal submitted by the cable and consumer electronics industries. They provide a mechanism to maintain the 'chain of custody' for encrypted programming and specify the types of DRM instructions that MVPDs can pass along to limit copying. In the cable context, the security module that subscribers obtain from their cable company performs the conditional access[23] and decryption functions and outputs a program stream that includes the DRM instructions. The 'plug and play' rules require the consumer equipment into which the security module is inserted, including television receivers and set-top boxes, to implement the DRM instructions.

The instructions range from 'copy never' for pay per view or video on demand programming to 'copy freely' for broadcast channels carried by the service. In essence, the FCC rules place limits on content providers' ability to restrict copying of programming distributed by MVPDs by restricting the nature of the DRM instructions that MVPDs may transmit.

On 4 November 2003 the FCC adopted the 'Broadcast Flag' rules intended to limit Internet retransmission of digital broadcast programming by requiring new DTV receivers to recognize a marker broadcasters can put on their digital content.[24] These rules did not prevent consumers from freely making copies of digital content for their own use (as they can now do with analogue content) on digital recorders. However, they were adopted based in part on the premise that widespread redistribution of broadcast programming without quality degradation would cut into broadcasters' advertising revenues, and the viability of digital over-the-air broadcasting might be threatened. Content producers might be reluctant to provide high-value programming to broadcasters because of fear of unlimited retransmission. As explained above, in May 2005, the US Court of Appeals for the DC Circuit found that the Commission did not have statutory authority to adopt the broadcast flag rules and overturned them.

COMPLETING THE DTV TRANSITION

This section begins by reviewing early legislation that specified the terms for completing the DTV transition. Next is a discussion of recent legislation that sets a hard deadline for completing the transition. The legislation also provides for a set-top box subsidy.

Early DTV Legislation

The 1997 legislation for the DTV transition set a deadline of 31 December 2006 for turning off analog broadcasting. This legislation has recently been superseded by new legislation, but is still of interest to students of alternative policies for managing the DTV transition. The 1997 legislation directed the Commission to extend the 2006 deadline for any requesting television station if one or more of the following three conditions applies.[25] First, the extension would have been available if, in the requesting station's television market, there is at least one station that is affiliated with one of the top four commercial broadcast networks and is not transmitting a DTV signal, and the Commission finds that such station or stations qualify for an extension of applicable Commission construction deadlines. Second, the extension would have been available if digital-to-analog converter technology is not generally available in the television market. Third, the extension would have been available if fewer than 85 per cent of television households in the market have access to DTV signals.

The 'access' criterion could have been met in three ways.[26] One means of access would have been subscription to 'a multichannel video programming distributor . . . that carries one of the digital television service programming channels of each of the television stations broadcasting such a channel in such market'. A second means of access would have been to have 'at least one television receiver capable of receiving the digital television service signals of the television stations licensed in such market'. The third means of access would have been to have at least one analog television receiver 'equipped with digital-to-analog converter technology capable of receiving the digital television service signals of the television stations licensed in such market.'

There are several points worth noting about the transition provisions of the 1997 legislation. First, it sets up the transition on a market-by-market basis, although the Commission did not define the term 'television market' in this context. A market-by-market transition may make sense for minimizing the loss of television service from the transition, but it makes less sense if the goal is reclaiming geographically contiguous blocks of spectrum. Second, the requirement was for reception or equipment capable of reception of digital television service and not the display of HDTV service. Thus a set that displays only SDTV quality video, even if the source is in HDTV, would have met the standard. This would have reduced costs to consumers because a SDTV set, at least initially, is likely to be significantly less costly than HDTV equipment.

With regard to the 85 per cent criterion, it is clear that the DTV receiver and digital-to-analog converter provisions did not, as of mid-2005, go very far toward meeting that threshold. As noted in the section on DTV equipment, only about 3.3 per cent of US television households as of the end of 2004 had either an integrated DTV receiver or a set-top box capable of receiving DTV signals and converting them to analog. In order to understand the role that the MVPD subscription provision would have played under the early DTV legislation, it is necessary to review the FCC's cable signal carriage regulations. This discussion is also relevant under the most recent DTV legislation because it describes the current FCC rules for retransmission of DTV signals via cable during the remainder of the transition. Pursuant to explicit provisions of the Communications Act of 1934, FCC regulations require cable operators to retransmit all local analog television signals, subject to certain limitations.[27] The signal carriage rules require that, every three years, each commercial television must elect either must-carry or retransmission consent status for their signals. Must-carry status means that, subject to the limitations referred to above, a cable operator is required to retransmit the station's signal without compensation and to provide it with certain channel positioning rights. Retransmission consent means that a cable operator is prohibited from carrying the signal without the consent of the station licensee. Compensation may be paid by either party as part of a retransmission consent agreement.

The statute also requires the FCC to issue signal carriage regulations for DTV signals. The Commission has resolved many of these issues. Stations transmitting in both analog and digital formats are not entitled to 'dual must-carry'. Currently their must-carry rights apply only to their analog signals. Digital-only stations (of which there are not many), however, have must-carry rights now. Digital must-carry rights apply to one video program stream, so stations that 'multicast' can only have one of their program streams transmitted pursuant to the must carry rules. Under current regulations, during the transition, 'a television station may choose must-carry or retransmission consent for its analog signal and retransmission consent for its digital signal'.[28] Thus, currently, for stations transmitting both in analog and digital modes, if the digital signal is retransmitted over cable, the carriage is pursuant to voluntary agreements between the stations and the cable operators. (Those stations transmitting only in digital mode have the right to elect must-carry or retransmission consent status for their digital signals.) A variety of stations transmitting in both modes have secured carriage of their digital signals under voluntary agreements.[29] The fact that a significant number of stations opt for must carry status for their analog signals suggests that not all local television stations will secure carriage for their digital signals via retransmission consent during the transition. This means that, in some and perhaps most markets, households would not be considered to have access to digital signals for purposes of the 85 per cent criterion based on their cable subscription.

DTV signal carriage rules have not yet been finalized for DBS. Moreover, unlike in the cable case, the DBS signal carriage rules do not require retransmission of all local analog television signals. DBS carriers are subject to the requirement to carry all local analog signals in a market if they choose to carry one signal pursuant to a compulsory copyright license. Additionally, unlike cable, there is no requirement that DBS customers receive the local signals. The subscriber may choose to receive them but at an additional charge. As with cable, then, some households would not be considered to have access to digital signals for purposes of the 85 per cent criterion based on their subscription to DBS.[30]

Recent DTV Legislation

Congress recently passed budget legislation with provisions addressing the DTV transition.[31] The legislation sets a 'hard date' of 17 February 2009 for turning off analog service. That is, analog stations would have to cease transmissions in all markets regardless of the penetration of DTV equipment in the markets. The legislation provides funding not to exceed $990 million for converter box subsidies, with a provision to increase that amount up to $1.5 billion if the agency administering the program submits a request to congress justifying the increase. The legislation sets a 28 January 2008 deadline for commencing the auction of spectrum to be recovered for commercial use.

Set-top box subsidy plans

Under the legislation, the National Telecommunications Information Administration (NTIA) would administer a program to give, upon request, up to two $40 coupons per household to be used towards the purchase of digital-to-analog converter boxes. NTIA can spend up to $160 million on administrative expenses. If the full $160 million is used for administrative expenses, $830 million of the $990 million initially authorized would remain to purchase 20.75 million converter boxes.

The idea of a set-top box subsidy has been widely examined. Issues under discussion included whether to subsidize all 'over the air' households or just those that meet a means test, what the means test should be, how much a subsidy would cost, how it should be financed (possibly from the revenues earned by auctioning spectrum now assigned for television), what equipment (basic digital to analog converters, fully integrated DTV receivers) should be eligible for the subsidy, how the program should be administered, and many others. The estimated cost of a subsidy program will vary significantly depending on the assumptions made about eligibility and the cost of a set-top box.

In Congressional testimony, the Government Accountability Office (GAO) provides cost estimates for two subsidy plans. The first option provides for a subsidy for one converter box per household with only analog OTA televisions. GAO found that 19 per cent of total US households (not television households) rely exclusively on OTA television.[32] The cost of providing each of these 20.8 million households with a converter box would be $1.04 billion dollars at $50 per converter. Another option GAO examines is a subsidy for one set-top box to every household that relies solely on OTA reception and has income less than 200 per cent of the poverty level. GAO estimates that about 9.3 million households (roughly 50 per cent of estimated total OTA households) would be eligible and that the plan would cost $463 million at $50 per converter box.[33]

The Consumers Union/Consumer Federation of America (CU/CFA) advocates more expansive subsidy plans. Specifically, they recommend that Congress 'should allocate funding based on the number of OTA-only sets and OTA-only households reflected by the higher end of estimates provided to date'. They suggest that the number should be at least 70 million.[34]

In the course of arriving at its own estimate, CU/CFA references estimates from other sources, including the National Association of Broadcasters (NAB) and the Consumer Electronics Association (CEA). The NAB estimates that 73 million analog television sets are not connected to an MVPD.[35] A subsidy program to provide converters for all such receivers, using the NAB estimate and Motorola's estimate of $50 per converter box[36] yields a total cost of purchasing converter boxes of $3.65 billion. But, according to the CEA, only 33.6 million television sets are actually used for OTA reception.[37] A subsidy plan that would provide converter boxes for only these television sets would cost of $1.68 billion.

Because the legislation provides for funding the subsidy program from auction revenues, and because there has also been other discussion of this option, it is worth looking at recent estimates of the value of the reclaimed spectrum to be auctioned. Reported estimates range from $10–28 billion, so auction revenue is likely to cover even the most expansive subsidy options.[38]

CONCLUSION

Recapping, policy makers in the US care about the transition to digital broadcast TV for three reasons. First, it provides more and better services for consumers. Second, it maintains a separate platform that can compete with cable and satellite. And last, the transition to digital broadcast TV could release a lot of spectrum for other valuable uses. Facilitating the digital transition is a high priority for the FCC and Congress. The major policy challenges include digital rights management and reclaiming spectrum quickly efficiently and equitably.

NOTES

1. The opinions expressed herein are those of the authors and do not necessarily represent the views of the Federal Communications Commission or any other member of its staff. The authors are grateful to John Williams for his helpful suggestions and comments.
2. The term 'DTV' is generally used to refer to terrestrial, over-the-air transmissions, although other distribution platforms, such as cable and DBS, also employ digital modulation for some or all of their transmissions. The particular transmission standard used in the United States for terrestrial DTV is the ATSC standard.
3. The upper 700 MHz band consists of UHF channels 60–69 on 746 MHz–806 MHz.
4. The lower 700 MHz band consists of UHF channels 52–59 on 698 MHz –746MHz.
5. The transition will also make additional spectrum available on channels 2–51 for additional DTV stations or possibly unlicensed cognitive radios.

6. Thomas Hazlett and Robert Munoz have estimated the social value of spectrum made available for commercial use. See, 'What Matters in Spectrum Allocation Design,' (AEI-Brookings Joint Center for Regulatory Studies Working Paper 04–16, August 2004). http://mason.gmu.edu/~thazlett/pubs/What%20really%20matters_hazlett.pdf
7. Fourth Report and Order in MM Docket 87–268. 11 FCC Rcd 17771 (1996).
8. Fifth Report and Order in MM Docket 87–268. 12 FCC Rcd 12809, 12850–51 (1997).
9. See Communications Act of 1934, as amended; 47 USC Section 336 and 309(j)(14).
10. Stations may also retain their analog licenses if at least one station in the market that is affiliated with one of the top four commercial broadcast networks is not transmitting a DTV signal (and the Commission finds that the station or stations in question qualify for an extension of applicable Commission construction deadlines) or if digital-to-analog converter technology is not generally available in the market.
11. The Bill (S. 1932) passed the Senate on 21 December 2005 and the House on 1 February 2006 and was signed by the President 8 February 2006.
12. FCC, July 2005.
13. Trade publication statistics, which some consider to be an overestimate, indicate that 108.2 of the 109.6 US television households are passed by cable. These figures, from Kagan Research LLC and A.C. Nielsen Media Research, respectively, are cited on the NCTA website. See www.ncta.com/Docs/PageContent.cfm?pageID=86 (visited 2 May 2005). The Commission's *Hearing Designation Order* in the proposed Echostar-DirecTv merger discusses the different sources of homes passed data and the methodological issues involved in calculating homes passed as a percentage of television households. At that time, the applicants, apparently using Kagan data, stated that over 96 per cent of television households were passed by cable and hence under 4 per cent were not passed by cable. Other sources and methodologies provided a range of homes not passed of 9.86–21.28 per cent. See *Hearing Designation Order* in CS Docket No. 01–348. 17 FCC Rcd 20559, 20611–12 (2002)
14. Data as of December 2004. National Cable & Telecommunications Association, Website visited 28 July 2005. http://www.ncta.com/Docs/PageContent.cfm?pageID=86. In general, the cable digital tiers consist of analog channels that have been digitized and compressed to conserve on cable system capacity.
15. Testimony of Kyle McSlarrow, President and CEO, National Cable & Telecom-munications Association, on The Digital Television Transition Act of 2005 [Staff Draft]. National Cable and Telecommunications Association Press Release 5/26/05. http://www.ncta.com/press/press.cfm?PRid=604&showArticles=ok; visited 16 August 2005.] In the overwhelming number of cases, the cable system is delivering the DTV signal con-verted to analog format.
16. 'Public Television and Cable Announce Major Digital Carriage Agreement.' National Cable and Telecommunications Association Press Release, 31 January 2005. http://www.ncta.com/press/press.cfm?PRid=573&showArticles=ok; visited 1 August 2005.
17. See Second Report and Order and First Order on Reconsideration in CS Docket No. 98–120, 20 FCC Rcd 4516 (2005).
18. See Second Report and Order in ET Docket No. 05–24 (released 8 November 2005).
19. RCA has announced that a 27 inch standard definition DTV receiver will be available in 2005 at a suggested retail price of under $300 http://tv.rca.com/en-US/01052005.html (visited 7 July 2005). See also Communications Daily, 2 May 2005, p. 7 (reporting a claim from Zoran Corporation (http://www.zoran.com/) that they can manufacture a set top converter box in quantity today for $50 with the capability to convert over-the-air DTV signals of local television stations to 'DVD-quality' pictures on an analog television set.) Zoran demonstrated their set top converter box at the FCC on 28 April 2005.
20. American Library Association, *et al*. v. FCC, No. 04–1037 (DC Circuit) decided 6 May 2005.
21. Second Report and Order and Second Further Notice of Proposed Rulemaking in CS Docket 97-80 and PP Docket 00–67. 18 FCC Rcd 20885 (2003).
22. FCC rules effectively exempt DBS services (whose equipment is widely available from independent retailers) from the requirement. See 47 CFR Section 76.1204. Cable

operators were required to offer the separate security modules as of 1 July 2000 and were required to stop deployment of new 'integrated' navigation devices on 1 January 2005. The deadline has been extended twice, most recently until 1 July 2007. See Second Report and Order in CS Docket 97-80 (released 17 March 2005).

23. That is, the module verifies that the receiving household is, in fact, eligible to view the programming in question.

24. Report and Order and Further Notice of Proposed Rulemaking in MB Docket 02–230. 18 FCC Rcd 23550 (2003).

25. Communications Act of 1934, as amended. 47 USC Section 309(j)(14)

26. The Commission has not issued rules implementing these criteria. See, FCC, Second DTV Periodic Review.

27. See 47 USC 614 and 615, esp. 614(b) and 615(b). Cable operators are not required to devote more than one third of their 'usable activated channels' to local commercial television stations, nor are they required to carry more than one affiliate of the same commercial broadcast network or to carry a station that substantially duplicates the signal of another local commercial television station carried by the cable system. Cable operators are also not required to carry stations that do not provide a good quality signal to the principal headend of the cable system.

28. First Report and Order and Further Notice of Proposed Rulemaking in CS Docket N. 98–120. 16 FCC Rcd 2598 (2001), para. 27. The Commission's digital must-carry proceeding remains open and it could adopt additional regulations regarding digital signal carriage during the transition.

29. See the discussion supra of the agreement between the Association of Public Television Stations and the NCTA.

30. Nevertheless, it appears clear that cable and DBS are destined to play a major role in the DTV transition. The most recent FCC 'Competition Report' indicates that, as of June 2004, 82.3 per cent of US television households subscribed either to cable (60.97 per cent) or DBS (21.36 per cent). Annual Assessment of the Status of Competition in the Market for the Delivery of Video Programming (2004 Video Competition Report) 20 FCC Rcd 2895 (2005), Table B-1. More recent data from the same sources used for the Competition Report show that, at the end of 2004, 82.4 per cent of US television households subscribed to cable or DBS. As noted above, however, not all DBS subscribers have access to retransmitted local television signals, and not all who have access choose to receive them. See Kagan Research, LLC Cable TV Investor, 31 March 2005, p. 3 (cable and DBS subscribers) and Television Bureau of Advertising, www.tvb.org/nav/build_frameset.asp?url=/rcentral/index.asp (visited 7 July 2005) (Nielsen figures on television households).

31. The Senate passed S. 1932 (Budget Reconciliation Bill Conference Report) on 21 December 2005 and the House passed H. Res. 653 (a resolution that agreed to the Senate amendments to the Conference Report on S. 1932) on 1 February 2006. S. 1932 was signed by the President 8 February 2006.

32. See Digital Broadcast Television Transition: Estimated Cost of Supporting Set-Top Boxes to Help Advance the DTV Transition. Statement of Mark L. Goldstein, Director Physical Infrastructure Issues, Government Accountability Office (17 February 2005).

33. Ibid. p. 14. Note that this corresponds to the cost of the subsidy program in the Senate discussion draft excluding the $5 million the draft provides for administrative costs.

34. See Consumers Union/Consumer Federation of America, 'Estimating Consumer Costs of a Federally Mandated Digital TV Transition: Consumer Survey Results', pp. 6–7 (29 June 2005). http://www.hearusnow.org/fileadmin/sitecontent/DTV_Survey_Report_Final_6-29-05.pdf

35. Comments filed at the FCC in MB Docket No. 04–210 by the National Association of Broadcasters and MSTV, at ii, 2. See FCC Media Bureau Staff Report on OTA Viewers, p. 11, ftn. 66. The 73 million total includes 28 million receivers in MVPD households that are not connected to the MVPD and 45 million receivers in 'over-the-air' households.

36. Motorola's estimated a cost of $50 per converter box by 1 January 2009. See *Communications Daily*, 13 July 2005. p. 3. Zoran, however, claims that they can manufacture boxes in quantity today at this price. See Note 19 above.

37. See www.ce.org/press_Room/press_release_detail.asp?id=10764 (visited 20 June 2005).
38. A study commissioned by Intel includes a $24 billion estimate, while a member of Congress cited the Congressional Budget Office as the source of a $10 billion estimate. (*Communications Daily*, 31 May 2005, p. 1). Qualcomm has funded a study by the Brattle Group that contains a $28 billion estimate (*Communications Daily*, 19 May 2005, p. 9). And the New America Foundation has published a lower bound estimate of $10 billion, 'Speeding the DTV Transition: Facts and Policy Options,' p. 3; www.newamerica.net/Download_Docs/pdfs/Doc_File_2389_1.pdf; (visited 20 June 2005)

3. Digital broadcasting in the developing world: a Latin American perspective

Hernan Galperin

INTRODUCTION

The broadcast industry is in the midst of its most important transformation since its beginnings almost a century ago. The basis for this transformation lies in the digitalization of the broadcast signals carried over the main transmission platforms: cable, satellite, and in the case of Latin America, the terrestrial broadcasting network. This is certainly not the first time that the television industry faces reorganization on a massive scale. But for the most part, past technological transformations have spurred evolutionary, not revolutionary change. In most cases, black-and-white TV sets were still able to pick up several colour TV signals. Analogue cable and satellite TV largely brought more of the same: branded one-way programming services. The transition to digital TV is different. It implies a complete re-tooling of the existing video production and distribution infrastructure, from studio cameras to transmission towers. It requires new mechanisms to compensate content creators and distributors in a world where perfect copies can be easily made and redistributed by viewers. And it raises multiple questions about the adequacy of the existing legal apparatus for the communications industry at large.

In a sense, the transition to digital TV is about a revolution long overdue. Compared with related sectors, the pace of technological change in the broadcasting industry over recent decades has been much slower. Digital technologies have revolutionized the telecommunications and information services industry, as well as the production of audiovisual content. Yet the continued utilization of analogue equipment in the transmission and reception of video programming has, until recently, sheltered the broadcast industry from the winds of change that have swept related sectors. The transition to digital TV removes this protection. Today, the same forces that have radically transformed the telecommunications and information services industries are beginning to revolutionize the structure of the broadcast sector.

While these transformations have been under way for some time in the developed world, Latin American nations have only recently started to set the terms for the transition to digital broadcasting. This has raised several interesting questions about the allocation of spectrum property rights, the adoption of digital broadcasting standards, and the transition strategy better suited for the countries in the region. Furthermore, there is much interest among regional policymakers in the opportunities offered by digital broadcasting to alleviate existing disparities in access to information and communication services, as well as for spurring growth and enhancing R&D capabilities in the local information technology sector. After a decade of reforms and sustained growth in the telecom sector, there is growing consensus about the need to extend reforms and spur innovation in the broadcasting sector.

This chapter examines the development of digital broadcasting in Latin America, with a particular focus on the ongoing debates about spectrum management reform, digital TV standards, and the opportunities associated with interactive TV as a low-cost platform for the delivery of information services. Because of the relatively low penetration of cable and satellite in most countries in the region (see Table 3.1), particular attention is given to the debates and transformations related to terrestrial broadcasting. The main argument is that the transition to digital broadcasting represents a unique opportunity to reformulate the existing industry structure based on limited competition between a handful of licensees, as well as to optimize spectrum utilization and promote innovative radio-based communication services at the local level. However, numerous political and institutional factors favour adaptation to existing industry arrangements over discontinuous change in the organization and legal arrangements of the sector, thus minimizing the opportunities for policymakers to seek reforms.

The chapter is organized as follows. The first section outlines the potential benefits and challenges for adoption associated with digital broadcasting in the developing world. The following section outlines the major developments related to digital broadcasting (particularly in the terrestrial TV sector) in three key Latin American markets: Argentina, Brazil and Mexico. Given the space limitations in this chapter, these examples are meant to illustrate the main factors at play in the adoption of digital broadcasting across the continent. The conclusion discusses the opportunities and challenges of different government strategies *vis-à-vis* standards-setting, the promotion of local electronics industries, and digital inclusion initiatives related to digital broadcasting.

Table 3.1 Information and communication technologies in Latin America in 2003 (selected countries)

	TV Sets (in millions)	TV Owners-hip (% HH)	Pay-TV (% HH)	Telephone subscribers per 100 inhabitants	Mobile subs. as % telephone subs.	PCs per 100 inhabitants	Internet subs. per 100 inhabitants
Argentina	10.6	98	54	39.6	45	8.2	3.9
Brazil	53.7	88	12	50.4	52	7.5	4.5
Chile	3.5	93	28	73.2	70	11.9	5.7
Colombia	8.0	98	4	34.1	41	4.9	1.6
Mexico	25.0	96	20	44.8	65	8.2	2.0
Peru	3.6	85	19	17.3	61	4.3	1.5
Venezuela	4.3	96	25	36.9	70	6.1	1.3

Sources: ITU, Zenith Media, The Economist Intelligence Unit.

DIGITAL BROADCASTING: OPPORTUNITIES AND CHALLENGES FOR LATIN AMERICA

In technical terms, digital TV offers a number of advantages over analogue broadcasting, such as increased bandwidth use efficiency, increased inter-operability with telecommunications and computer industry hardware and applications, and increased flexibility for the provision of services other than traditional audiovisual programming (Cave, 1997; Owen, 1999). For existing operators, it represents an opportunity to expand the range of services, lower transmission costs on a per channel basis, and tap into new revenue streams. Yet the erosion of bandwidth bottlenecks has renewed questions about existing spectrum and broadcast licensing policies (for example, Hazlett and Spitzer, 2000; Sunstein, 2000). This has particular relevance for Latin America, which has historically followed the US commercial licensing model based on the granting of a small number of high-power licences to private firms in return for ill-defined (and politically-laden) programming obligations.[1]

Some of the advantages associated with digital broadcasting are particularly relevant to developing nations, where communications infrastructure tends to be generally underdeveloped and distributed unequally in favour of wealthy urban centres. First, digital broadcasting enables optimization in the use of valuable radio frequencies below 3 GHz. Data manipulation techniques (for example, MPEG) allow the compression of digitized audiovisual signals so that they can be transmitted more efficiently (that is, utilizing less bandwidth) than analogue signals. Moreover, digital broadcasting requires less spacing between channels, therefore allowing for the use of so-called 'taboo channels', largely unused in analogue transmissions. The increased spectrum efficiency of digital broadcasting thus opens multiple opportunities for changing the structure of property rights for prime frequencies, allowing not only the licensing of more radio services but also experimentation with non-exclusive licensing regimes that contemplate frequency sharing among different types of users. Making more of these desirable frequencies available to a more diverse set of market actors would help fuel the continued growth of wireless communication infrastructure in developing regions. This is critical since, as the ITU (2004) notes, progress in telephony penetration in the developing world over the past decade has been largely the result of growth in mobile telephony (see Table 3.1). In fact, of the countries where penetration of mobile telephony first surpassed that of fixed lines, most were in Africa and Latin America.

Second, digital broadcasting may offer a cost-effective gateway into the Information Society for those currently underserved by existing operators or unable to afford the equipment and service costs involved in more conventional, PC-based Internet connections. This possibility has garnered significant attention from policymakers preoccupied with the so-called 'digital divide', particularly in countries with relatively low internet penetration.[2] As Table 3.1 reveals, the contrast is apparent in the case of Latin America: while most households have at least one TV set, few of them have computers or Internet connections. The idea that digital broadcasting could significantly alleviate existing inequalities in access to advanced information services is none the less problematic for a number of reasons. To begin with, household Internet subscription figures grossly underestimate actual Internet adoption rates since most users in Latin American (and other developing regions) connect through shared access centres (either privately-run cybercafés or public telecentres). There are also inherent limitations related to interactive services over terrestrial broadcast networks, particularly in cases where most households lack fixed-line return connections (which as Table 3.1 reveals is the case in Latin America). However, as discussed below, this has not deterred governments in the region from touting digital broadcasting as an important digital inclusion tool.

Third, the migration to digital broadcasting offers opportunities for proactive governments to sponsor the local development of digital broadcasting and related technologies through the standards selection process. National (or regional) policies in support of domestic technology firms are a highly contentious proposition which none the less has a long and established tradition in the broadcast industry, notably in the case of colour TV standards (Crane, 1979) and the development of analogue HDTV systems (Galperin, 2004). Much has changed today since the days of PAL vs. NTSC vs. SECAM.

Among other things, the national allegiances of the main competing digital TV standards (ATSC, DVB, and ISDB) are much less well-defined.[3] None the less there has been little abatement of government efforts to promote 'national' (or in the case of Europe 'regional') systems, notably by mandating use within domestic markets and by aiding companies in diplomatic efforts to gain adoption in other nations (Hart and Prakash, 1997). Overall, the role of national governments in shaping competition in the broadcast technology market, if less hostile than in the analogue world, has hardly subsided. As discussed below, Brazil is a case in point.

The opportunities discussed above explain why several Latin American governments, despite a number of other more pressing priorities, have engaged in concerted efforts to formulate and implement a plan for the transition to digital broadcasting. Yet the experience of developed nations also reveals a number of challenges in the migration from analogue to digital broadcasting. Among them: co-ordination problems between the different market actors (network operators, programmers, equipment manufacturers, and so on); market failures stemming from inadequate allocation of spectrum property rights that create misalignments between private incentives and welfare-maximizing transition paths; gridlocks that result from the ability of key stakeholders to slow the transition through political action in different policy arenas (Geller, 1998; Levy, 1999). Generally speaking, the experience from nations that have first embarked in the transition process reveals two distinct approaches to address these challenges.[4]

The first approach is best embodied by the US experience, and its central premise is the extension of the existing industry arrangements into the digital broadcasting era. This approach is revealed in the policy choices made with respect to spectrum allocation (for example, the much criticized allocation of a second 6 MHz channel to existing licensees for the launch of digital services) and consumer equipment specifications (for example, the digital tuner mandate), which have largely benefited incumbent local broadcasters at the expense of other stakeholders. The second approach is best exemplified by the UK experience, and its central premise is the reallocation of spectrum rights and the greater government involvement in managing the transition and addressing potential co-ordination problems. This is best exemplified in the attempt to increase competition by licensing new digital services, in the formulation of a detailed digital switchover plan, and in the extensive regulatory reforms that have accompanied the transition. Table 3.2 provides a concise summary of the key differences in the two approaches.

Policymakers in Latin America thus faced multiple challenges and an array of options with respect to the setting of standards, the licensing of

Table 3.2 Two approaches to the digital TV transition

	US	UK
Co-ordination	Market-driven, except for transmission standard (ATSC) and timetable for launch of services	EU-sanctioned standard (DVB), government more actively co-ordinates transition process
Licensing	Incumbent broadcasters receive additional licence	Digital licences available to new operators
Regulatory model	Based on existing analogue broadcast regime	New regime based on separation between infrastructure and services
Goals	HDTV services replicate analogue broadcast model	Spectrum use optimization, increased competition, interactive services

services, the managing of radio frequencies, and generally the creation of the appropriate conditions for the adoption of digital broadcasting. As we shall see below, not all of them have made the same choices, and the choices made have been driven as much by international forces as by – oftentimes changing – local circumstances. As noted, given space limitations and the sheer number of nations in the region, the analysis focuses on three relevant cases: Brazil, Argentina and Mexico. Their selection was based on the size of their internal markets (and hence regional influence) as well as on the degree to which the introduction of digital broadcasting services has been part of the national agenda.

A REGIONAL OVERVIEW: BRAZIL, MEXICO AND ARGENTINA

Brazil

Brazil provides the more interesting case study in Latin America as it is the only nation in the region where digital broadcasting has been a priority in the national government agenda. It is also the largest regional market, and thus exercises considerable political clout over its neighbours. Lastly, the organization of the analogue TV market is common to many Latin American nations: a dominant national network (Rede Globo) with a handful of smaller private competitors (SBT, Bandeirantes, Record, RedeTV) and a collection of even smaller public and not-for-profit broadcasters with limited national reach, the most important being TVE (based in Rio de Janeiro) and TV Cultural (based in São Paulo).

The debates about digital broadcasting in Brazil date back to 1994, when the main broadcast industry trade groups – the Broadcast Engineers Society (SET in Portuguese) and the Brazilian Association of Radio and Television Broadcasters (ABERT in Portuguese) – formed a joint technical committee to study the implementation of digital TV. At the time, the main concern was to evaluate which of the competing standards available would be best suited for adoption. In 1998, the newly created telecommunications regulator ANATEL took interest in the matter and called for a round of lab and field tests comparing the different systems available (ATSC, DVB, and later ISDB).[5]

These tests were conducted between October 1999 and April 2000, and were closely monitored by the industry and stakeholders of the competing systems. To the surprise of many, the results give a slight advantage to the Japanese system ISDB, even when at the time this standard had not been deployed in any market.[6] The selection process is further complicated by

US-backed efforts within CITEL (Interamerican Telecommunications Commission, an organ of the Organization of American States) to promote adoption of a single standard in the continent. As the political and economic stakes escalated, in 2001 ANATEL decides to launch a public consultation on the transition to digital broadcasting. The goal of the consultation is to create a focal point for discussions about a range of issues related to digital broadcasting, from the question of which standard to adopt to other issues such as the licensing of digital services, spectrum reallocation, and so forth. By then, it is clear that the complexity of the issues at stake went beyond ANATEL's jurisdiction, and that the Cardoso government was opting to defer the key choices to the next administration.

The 2002 general elections mark a major shift in the Brazilian digital broadcast strategy. The new administration of President Lula takes a more strategic approach to the transition, making digital broadcasting a central component of two larger administration goals: first, to revitalize the Brazilian consumer electronics industry, and second, to address the significant inequalities in access to advanced information and communication services among the Brazilian population (Bolaño and Vieira, 2004). These new guiding principles will affect policy choices on a range of issues, most notably the decision about standards for digital TV.

In November 2003, a presidential decree created the so-called Brazilian Digital TV System (SBTVD). While not explicitly rejecting the adoption of the existing standards, it is clear that the new administration favours the development of a new system by a consortium of domestic research centres and technology firms. Interestingly, digital TV is also discussed as a possible flagship initiative for high-tech co-operation with other large developing nations such as China and India. According to the SBTVD framework, the government intends to adopt a standard that is appropriate for the existing infrastructure of the Brazilian broadcast industry, the levels of income of the population, and the installed capacity of the domestic electronics sector. So far, the government has committed about US$20 million for the initial phase of the SBTVD, financed through the Telecommunications Technology Development Fund (FUNTTEL in Portuguese).[7] This represents a small fraction of the US$1.6 billion that ABERT estimates it will cost to digitize the Brazilian TV industry (not including consumer equipment costs).

The Lula administration has opted for a risky strategy in the development of a local standard for digital TV. There is little doubt that the market conditions in Brazil are very different from those in the developed nations where the existing standards originated. In this sense, a low-cost system that could integrate video programming with other information and communication services could be considered more appropriate than one developed around HDTV services. The country is also quite unique in the region (along with Mexico) because it is host to a sizeable consumer electronics sector in the Manaus Industrial Pole – mainly dedicated to TV set manufacturing but with firms also involved in electronics components and transmission equipment – with revenues of US$4.78 billion in 2004.

However, the development of a novel digital TV standard is a costly long-term proposition which, given the limited size of the Brazilian market (about 53 million TV sets), could prevent consumers from taking advantage of the economies of scale and innovation curve of the existing alternatives. The failed experience of the European (EUREKA) and Japanese (Hi-Vision MUSE) initiatives in the development of analogue HDTV systems, along with the mixed Brazilian experience in nurturing a local computer industry through the creation of a 'greenhouse' environment (Evans, 1995), calls for careful reflection on the cost-effectiveness and expected spillovers generated by the SBTVD initiative.

Other critical issues related to the licensing of digital services and spectrum reallocation are still under consideration by ANATEL as well as the Ministry of Communications. Yet from the documents produced so far there are strong indications that Brazil is prepared to follow the US approach by allocating an extra channel to all existing incumbents for simulcasting during the transition period. For example, ANATEL has already began channel allotment plans to accommodate a second 6 MHz channel for existing licensees in Brazil's 145 local broadcast markets, which would leave only a handful of channels available for new services.[8] At the same time, the broadcast trade groups have suggested that the additional channel be returned only after 90 per cent of the Brazilian households have acquired a digital decoder – which the US experience indicates would take several decades in the absence of additional adoption incentives for broadcasters and viewers.

Mexico

Mexico is another case study that illustrates the organization of the Latin American TV industry, where a virtual duopoly exists between Televisa (the dominant operator with four national channels) and TV Azteca (two national channels). Yet the case is also peculiar for a number of reasons. First, because of Mexico's vast border with the US, which requires attention to channel co-ordination and signal spillovers. Second, as a result of consecutive trade agreements (notably NAFTA), there is a large presence of consumer electronic manufacturers on the northern part of the country that take advantage of lower labour costs to supply the US market (part of the so-called maquiladora industry). Third, Mexico is more integrated with the US than with the rest of the continent in economic and to some extent political terms. These facts differentiate Mexico from other cases in Latin America, and have created incentives for a much closer alignment with the transition approach of its northern neighbour.

The implementation of digital broadcasting got under way in Mexico in July 1999 with the creation of the Consultative Committee for Digital Broadcasting Technologies (CCTDR), composed of representatives from government and industry. At the same time, channels are allocated for experimentation with different broadcast standards. Though it was widely expected that Mexico would choose the ATSC system to facilitate frequency co-ordination and take advantage of economies of scale in set manufacturing for the NAFTA economic zone, extensive experiments are conducted with the three competing standards. In October 2003 the CCTDR presented its final report, which to no surprise recommended adoption of the ATSC system, as well as the allocation of an additional channel to incumbent licensees to simulcast their existing programming during the transition period and the establishment of a six-phase adoption plan.

Based on these recommendations, in July 2004 the government established the terms for the transition to digital TV in Mexico. Following the CCTDR recommendation, it adopted the ATSC system, a decision that is justified not so much on technical grounds but rather on the requirements for channel co-ordination with the US along the border as well as on the market expansion opportunities that digital broadcasting presents to the Mexico-based consumer electronics industry. Following the US approach, the government allocated a second channel to incumbent broadcasters for the simulcasting of existing programming to ensure minimal service disruption during the transition phase. Yet, contrary to the US case, it requires that digital signals be in a higher-definition format (either HDTV or EDTV).[9]

The government established an 18-year plan for the implementation of digital TV divided in six phases, starting with the launch of digital services in the main cities (Mexico DF, Guadalajara, Monterrey, Mexicali and Tijuana) by the end of 2006 and concluding with the full-blown launch of digital services across the Mexican territory by the end of 2021. Interestingly, the plan does not establish a precise date for the switch-off of analogue transmissions nor for the associated return of the analogue channels. Also missing is any discussion about promoting adoption of digital receivers, which is problematic in a country where the vast majority depends on terrestrial broadcasting and income levels are relatively low. Finally, the plan allows broadcasters to provide telecommunications services over their digital channels. This represents a valuable extension of their property rights over their allocated frequencies, since HDTV and in particular EDTV signals can be transmitted in less than 6 MHz, leaving significant room for the development of new radio-based services.

Argentina

Much like in the rest of Latin America, the implementation of digital broadcasting in Argentina has been driven largely by the private sector, notably commercial broadcasters and equipment manufacturers. Compared with that of Brazil and Mexico, the Argentine television industry is more fragmented, with three private competitors at the national level (Artear, Telefé and AzulTV), several regional players, a public service broadcaster (Argentina Televisora Color) with a small audience but a large national reach, and a vibrant independent production sector.

The debate about digital broadcasting began in July 1997, when upon request from the major trade groups the Menem administration created a commission to study and advise the government on the transition to digital broadcasting in Argentina. Sponsored by the Secretary of Communications, the commission is composed of representatives from the various trade groups, research institutions, professional societies, consumer defence groups, the telecommunications regulator (CNC) and the broadcast regulator (COMFER).[10] Shortly thereafter, the government authorized the main broadcast networks to begin experimental digital broadcast operations for a period of three years.

One of the main goals of these experimental operations was to allow testing of the different digital TV systems available. Argentina, along with neighbouring Paraguay and Uruguay, had previously opted for a unique colour TV standard (called PAL-N) that represented an adaptation of the European PAL system to the 6 MHz channel scheme dictated by the ITU for the Americas region (most other nations in the region opted for the American NTSC, which was originally designed for 6 MHz, while Brazil created yet another PAL adaptation called PAL-M). This effectively created a small isolated market, rising pricing and preventing technology transfers, while protecting weak local firms. The general consensus was that it was important to avoid a repeat of such policies.

Yet testing of the different systems ended abruptly in October 1998 when the Secretary of Communications adopted the controversial Resolution 2357/98, which imposed the ATSC system for digital TV in Argentina. According to the Secretary, the decision was based on four fundamentals:

(a) the fact that the Argentine channel scheme is similar to the one utilized in the US but different from that in the EU (8 MHz channels);
(b) the need to differentiate terrestrial services from cable and satellite competitors by offering HDTV services (which according to the Resolution are contemplated in the ATSC system but not in the DVB standard);

(c) the immediate availability of transmission and reception equipment for the ATSC system; and

(d) the opportunities for local electronic firms as well as programmers in the creation of a ATSC-based digital TV market in the Americas.

The Resolution raised much controversy for a number of reasons. To begin with, it adopted the ATSC system without thorough testing of alternative standards and transparent decision-making procedures. The succinct technical rationale presented is also questioned, particularly by advocates of the DVB system.[11] Moreover, the neighbouring government of Brazil demands an explanation for the lack of consultation as required by technical co-ordination agreements contained in the MERCOSUR framework. After the 1999 general elections, the new administration of President De la Rúa declared its intention to revise the controversial resolution, arguing that the decision was taken 'under erroneous assumptions, without technical testing or coordination with Brazil, and under false assumptions about the European DVB norm' (cited in Albornoz *et al.*, 1999). However, in the midst of the tumultuous economic and political crisis that culminated with the resignation of President De la Rúa in December 2001, the Resolution was never formally repealed.

Currently, digital broadcasting in Argentina is advancing at a very slow pace. The unprecedented economic crisis of 2001 left most operators cash-strapped, particularly in hard currency. Without the industry initiative, the transition has been moved onto the policy backburner. Yet is clear that political alignments have changed significantly with the new Kirchner administration. While Argentina is still formally listed as an ATSC market, the government has made clear that future steps will be guided by a more regional perspective, and has been closely monitoring the Brazilian efforts to develop a digital TV system suited for Latin America. In this respect, the new government has already voiced its support for what it called a 'non-aligned' digital TV standard, as part of a broader effort to foster regional co-operation and nurture domestic R&D in technology-intensive industries.

CONCLUSION

Developing nations face a myriad of challenges associated with the development of the communications infrastructure necessary for effective participation in the Information Society. In this context, the modernization of the terrestrial broadcast network, a cost-effective information transmission platform that is uniquely pervasive in the developing world, becomes most relevant. Yet is clear that the adoption of digital broadcasting will not be driven by the same goals as in the developed world. Policymakers in Latin America and elsewhere are hardly enticed by the prospect of delivering crisper images of telenovellas and soccer matches to the public, nor the industry is willing to invest in a costly technological transition that has proven elusive to generate profits in the more developed markets (Brown and Picard, 2005). Rather, the adoption of digital broadcasting in these nations is driven by three policy goals: spectrum management reforms, industrial policies and digital inclusion initiatives.

With respect to the first, the evidence from Latin America is not encouraging. The policy choices made so far in Brazil and Mexico reveal a penchant for the US approach that transfers valuable spectrum property rights to incumbent broadcasters but creates few incentives for these firms to speed the transition and return channels to the government for reallocation. Moreover, this approach would practically foreclose new entry in broadcast markets with a dismal pluralism record and a notorious history of quasi-monopoly behaviour by operators with close government ties.[12] New wireless technologies have allowed millions in Latin America to have a telephone for the very first time, and have helped bring Internet service to remote locations ignored by traditional telecommunications operators. A transition strategy aimed at optimizing spectrum use could open the floodgates for a host of new high- and low-power radio services, both public and private. Yet government strategies *vis-à-vis* digital broadcasting in Latin America seem timidly guided by the preservation of an antiquated and highly inefficient structure of spectrum rights.

The Latin American experience also reveals that choosing a technical standard for digital TV transmissions has been central to the debates, and that recent political changes have swung the balance in favour of the co-operative development of a system better suited to the social and economic realities of the continent. Interestingly, the issue is now recognized not as a purely technical question, but as one that touches on a range of national initiatives in the area of information and communication technologies. There are a number of reasons why this should be so, since the choice of standard will impact on local R&D capacity, on the ultimate cost of the transition (including foreign royalty payments), on participation in international standards-setting organizations, and possibly on the location of foreign direct investments in the consumer electronics sector. Yet the real question is whether the benefits of local technology development will outweigh possible delays and increased adoption costs. The evidence in related technology sectors points to adoption rather than production as the better strategy for less developed nations, particularly if the latter results in market isolation or attracts low-skill investments (Pohjola, 2001). This calls for careful evaluation before committing scarce government resources to well-intentioned initiatives that might yield too little, too late for countries in the region.

NOTES

1. There are only a handful of countries in Latin America with a history of strong public service broadcasters comparable to those in Europe and Asia. For a discussion see Fox (1988).
2. This in fact includes some European nations. See for example the references to digital TV in the eEurope 2005 Action Plan (EC, 2002).
3. The ATSC system was developed in the US by the members of the FCC-sponsored HDTV Grand Alliance, but includes two large European manufacturers (Thomson and Philips) and Zenith, which in 1999 became a wholly-owned subsidiary of the South Korean electronics giant LGE. The DVB system was developed by the EU-sponsored Digital Video Broadcasting consortium, but many American and Japanese firms are DVB members. ISDB, developed by a consortium of Japanese electronic manufacturers after NHK abandoned the Hi-Vision MUSE project, is the only system with a clear-cut national origin. Also, China is reportedly working on development of a local digital TV standard called DMB-T, but the system has yet to be released.
4. For a broader discussion see Galperin (2004).
5. It is important to note that according to the Brazilian legislation, ANATEL has jurisdiction over broadcast transmissions but not the licensing or regulation of broadcast services, which falls under the Ministry of Communications.
6. For a detailed discussion of the test results see CPqD (2001).
7. FUNTTEL is financed through a 0.5 per cent tax on the revenues of telecommunications operators and is managed by the Ministry of Communications.
8. For example, see Anatel Resolution No. 291, 13 February 2002.
9. EDTV (Enhanced Definition TV) is a format comparable in quality to the DVD signal. In the US, broadcasters were not required to broadcast in either HDTV or EDTV on their digital channels.
10. Under the current legislation the COMFER is responsible for the licensing of broadcasters. Formally independent from the Executive power, it has none the less been intervened by successive governments and effectively functions as a branch of the Executive.
11. While not contemplated in its original release for implementation in the EU, the DVB system was in fact designed to support channel schemes other than 8 MHz as well as HDTV services. For example, the DVB implementation in Australia supports 7 MHz channels and HDTV services.
12. Most obviously Televisa in Mexico and Rede Globo in Brazil (see Sinclair, 2002).

REFERENCES

Albornoz, Luis, Pablo Hernández and Glenn Postolski (1999) *La Televisión Digital en la Argentina: Aproximaciones a un Proceso Incipiente*, presented at the XXII Congresso Brasileiro de Ciências da Comunicação, Rio de Janeiro, Brazil.
Bolaño, César and Vinicius Vieira (2004) 'TV Digital no Brasil e no Mundo: Estado da Arte', *Revista de Economía Política de las Tecnologías de la Información y Comunicación*, 6(2), available at eptic.com.br
Brown, Allan and Robert Picard (2005) *Digital Terrestrial Television in Europe*, London: Lawrence Erlbaum.

Cave, Martin (1997) 'Regulating digital TV in a convergent world', *Telecommunications Policy*, **21**(7), 575–596.

CPqD (2001) Relatório Integrador dos Aspectos Técnicos e Mercadológicos da Televisão Digital, available at http://sbtvd.cpqd.com.br

Crane, Rhonda J. (1979) *The Politics of International Standards: France and the Color TV War*, Norwood, NJ: Ablex.

European Commission (2002) *eEurope 2005: An Information Society for All*, COM(2002) 263 final.

Evans, Peter (1995) *Embedded Autonomy: States and Industrial Transformation*, Princeton: Princeton University Press.

Fox, Elizabeth (1988) *Media and Politics in Latin America: The Struggle for Democracy*, London: Sage.

Galperin, Hernan (2004) *New TV, Old Politics: The Transition to Digital TV in the US and Britain*, Cambridge: Cambridge University Press.

Geller, Henry (1998) 'Public interest regulation in the digital TV era', *Cardozo Arts & Entertainment Law Journal*, **16**(2–3), 341–368.

Hart, Jeffrey A. and Aseem Prakash (1997) 'Strategic trade and investment policies: implications for the study of international political economy', *The World Economy*, **20**(4), 457–476.

Hazlett, Thomas W. and Matthew L. Spitzer (2000) 'Digital television and the quid pro quo', *Business and Politics*, **2**(2), 115–159.

International Telecommunications Union (2004) *The Portable Internet*, Geneva: ITU.

Levy, David A. (1999) *Europe's Digital Revolution: Broadcasting Regulation, the EU and the Nation State*, New York: Routledge.

Owen, Bruce M. (1999) *The Internet Challenge to Television*, Cambridge, MA: Harvard University Press.

Pohjola, Matti (2001) *Information Technology, Productivity, and Economic Growth*, Oxford: Oxford University Press.

Sinclair, John (2002) 'Mexico and Brazil: the aging dynasties', in Elizabeth Fox and Silvio Waisbord (eds), *Latin Politics, Global Media*, Austin: University of Texas Press.

Sunstein, Cass R (2000) 'Television and the public interest', *California Law Review*, **88**, 499–564.

4. DTT and digital convergence: a European policy perspective

Gilles Fontaine and Gerard Pogorel

Television policy in the European Union has been designed to conform to its overall aims of promoting European integration and social objectives. In its Sacchi Decision in 1974,[1] the European Court of Justice provided an illustration of the EC effort at abolishing national barriers to the free movement of goods and services, in stating that a television signal is considered a provision of services under Articles 59 and 60 of the Treaty of Rome. The EU's landmark initiative in television policy is the 'Television without Frontiers' Directive (EC 1989). It provides ground for the Community coordination of national legislation in the following areas covering fundamental principles applicable to the audiovisual industry: laws applicable to television broadcasts, promotion of the production and distribution of European works, access of the public to major (sports) events, television advertising and sponsorship, protection of minors, and right of reply.

Member States are required to conform or harmonise their national legislation to standards or criteria laid down in the text of the directive. This line of action has also led to audiovisual industry subsidy programmes, such as the European Commission's successive MEDIA programmes.

The review of the directive was launched in 2002 (EC 2002a), in the context of major shifts in technologies, increasing competitive pressure in the commercial TV sector, and concerns raised by the increase in the number of platforms and the multiplication of channels in the context of slow growth of overall resources available to the industry (see Perruci and Richeri, 2003). In a resolution adopted on 4 September 2003, the European Parliament called for a complete overhaul of the TVWF Directive that would take into account technological developments and changes in the structure of the audiovisual market. The Parliament wants the directive to be re-submitted in the form of a framework package which would bring together the current directive's underlying principles, the e-commerce directive and the directive co-ordinating certain copyright rules applicable to broadcasting and retransmission.

In conjunction with this industry-specific policy, the EU Action Plan eEurope 2005 has engaged in an effort at deriving the benefits from all

digital technologies and their convergence. Convergence is about providing together services that were previously distinct. It is made possible
by the more widely available usage of digitisation of information in areas
of technology and industries previously separated. The EU Action Plan
eEurope 2005 has also engaged in a specific effort concerning Digital
Television at large, and digital terrestrial television (henceforth DTT) in
particular. As stated in the 'Transition from analogue to digital broadcasting' Communication (EC2003a): 'Replacing analogue broadcasting with a
system based on digital techniques presents huge advantages in terms of
more efficient spectrum usage and increased transmission possibilities;
these will lead to new services, wider consumer choice and enhanced
competition.'

Since digital TV was introduced in 1996 in Europe, first on satellite
and soon after on cable and terrestrial networks, the European television
landscape has shown a wide variety of market situations. Approaches to
convergence have apparently diverged, at least in their initial stage.

A special feature of television, lies in this one very obvious, unique, statistical fact: PCs, according to the best medium term forecasts, will only be
present in at most 50 per cent of households, and 3G mobile broadband
telephony is expected to have about the same diffusion rate in the next 5, 6
or 7 years (World Economic Forum, 2005). Only television is already
present in nearly all households, the number of units itself being on the rise.
This is a fascinating reality for businesses, companies, marketers, and for
governments and regulators. This universal presence of television is highly
significant for information society services. Whereas their diffusion is
limited through the channel of PCs, and is also limited even on fairly optimistic assumptions regarding the diffusion of 3G, TV offers apparently the
most widely available platform for spread of such services. TV is now even
transforming itself, from a one-way diffusion channel to a potentially interactive channel. Digital interactive TV makes possible the creation of new,
TV-platform based, services, similar to those available on PCs or 3G mobile
terminals. It is against that background that we will now review the patterns
and evolutionary trends of television in Europe.

BACKGROUND ON TERRESTRIAL TV IN EUROPE

Terrestrial Broadcasting: from Number one to Residual Network

The terrestrial network, the veteran carrier of televised services, was long
the main, if not the only, platform available for broadcasting television in
the vast majority of European countries.

In most of these countries, through public radio and TV services, the state took charge of building the transmission infrastructure needed to cover the national territory. Only a handful of 'smaller' countries (chiefly Benelux and Switzerland) opted early on for cable, for both technical and cultural reasons (problems of available frequencies at the borders, population density, multilingualism and, for Switzerland, the country's very mountainous terrain).

Elsewhere, the subsequent introduction of first cable and then satellite left the analogue terrestrial network the time to become firmly ensconced in the TV landscape. Despite the new platforms' advantages, terrestrial broadcasting is still the predominant mode of TV reception for the majority of European households.

The other broadcasting platforms have nevertheless managed to attract a growing share of viewers in a great many countries, in part since the launch of digital television. Terrestrial transmission's share of viewership is therefore dropping steadily in all European countries as it loses out to the alternatives.

With the exception of Germany, terrestrial transmission is the largest platform in all of Europe's larger countries: the UK, France, Italy and Spain as well as Russia. It has the advantage of covering vast territories with a relatively dispersed population more easily and less expensively than other broadcasting technologies. Conversely, terrestrial is marginal in almost all of the smaller European nations, namely Benelux, Switzerland and Scandinavia. Many of these countries are enclaves within Europe, surrounded by larger countries and having either more than one national language (Switzerland, Belgium, Luxembourg), or a language that is little spoken outside their national borders (Scandinavia). Because of this, these countries were quick to opt for cable, and later satellite, whenever technically and economically possible, to distribute the channels developed inside their borders. This is because analogue terrestrial TV cannot carry a large number of channels. Opting for cable has made it possible to distribute a vast array of channels in those countries where it has not been feasible to develop a broad national offer on terrestrial broadcasting which has therefore been relegated to secondary status in a great many of Europe's smaller countries.

Terrestrial's migration to digital could, however, alter the situation:

- It could reverse the trend in countries where terrestrial TV still dominates, but is steadily losing a portion of its viewers to other platforms.
- In countries where it is already secondary, conversely, terrestrial TV may enjoy a revival of interest among viewers.

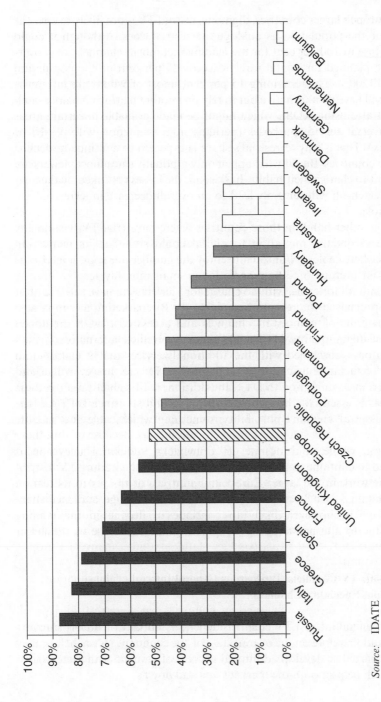

Source: IDATE

Figure 4.1 European households receiving TV services via the terrestrial network only, at the end of 2003 (percentage of TV households in selected countries)

In Europe's larger countries, digital terrestrial TV is not likely to cover as much of the population as analogue terrestrial does. Although it could reach close to 100 per cent for the leading incumbent channels, the coverage rate planned for digital rarely exceeds 85 per cent of the population (IDATE 2003). The remaining 15 per cent, most of whom are in remote areas, will have no alternative but to rely on another platform, mainly satellite, but also using ADSL which might be made available in certain areas for universal service purposes (perhaps in combination with Wi-Fi or Wi-Max). This is why terrestrial's share is expected to continue to decline in these countries. But, it will happen only gradually since most viewers are reluctant to change their habits. In the end, the losses provoked by the analogue switch-off will certainly lead to an overall decrease in terrestrial TV as a whole.

On the other hand, in those countries where terrestrial TV has already become secondary, migration to digital could stimulate its revival. In nations where cable and/or satellite enjoy the number one spot, households receive an average of 30 channels for a small monthly fee (around 10 euros a month for cable, with no charge for viewers who have satellite after having purchased the reception equipment). By marketing a comparable offer in terms of number of channels and access cost, with the added bonus of digital quality, DTT could become a credible alternative. Cable's digitisation – combined with the additional services that it enables and viewers' reluctance to change – and satellite TV are, however, likely to prevent a mass migration to DTT. In addition, DTT's chief target in these countries is essentially households with a second set, since DTT has the major asset of enabling portability, something which cable and satellite do not.

On the whole, and despite its drawbacks, terrestrial television is expected to continue to play a major role, either as the leading TV distribution network in the larger European countries, or thanks to its feature of portability, in a lesser role in the others. Its average audience share in Europe is, however, likely to continue to decline in the coming years, due to alternative networks' growing audience base in the larger countries.

Free-to-air TV Channels: Prosperous General Interest Channels, Precarious Speciality Channels

Private general interest TV channels depend essentially on advertising revenues, and therefore on the overall state of the economy. Nevertheless, they have managed to develop additional sources of revenue with merchandising, which help smooth the market's ups and downs.

Public channels are contending with problems at several levels:

- they are struggling clearly to define their mission of being general interest;
- alternatively they are being blocked in their bid to diversify to new commercial sectors; and
- controls on public spending limit their resources.

Speciality channels like news or sports depend directly on the economic environment of the leading multi-channel service operators. In most markets, the existence of several competing cable and/or satellite offers:

- has led to poorer health for multi-channel service operators, which are cutting the licensing fees paid to speciality channels; and
- yet, the creation of competing channel packages generates a supply which far exceeds potential demand.

In the US, all of the country's pay-TV platforms, both cable and satellite, broadcast the leading speciality channels. The result is a healthy dynamic: channels enjoy a very large audience; advertising revenues now account for a sizeable share of their revenues, and the licensing fees paid by platform operators can be cut, allowing them either to earn higher margins or to expand supply further. This scheme nevertheless requires that these platform operators distinguish themselves from competitors not by their channels but through prices or services.

The European market appears to be going the way of the American market. The merger of satellite services in Italy and Spain is leading to the creation of a single, dominant operator (such as BSkyB on the British market). A few exceptions remain, however, notably in France. This new landscape is likely to lead to a drop in the number of speciality channels available.

Transmission operators need new growth strategies. Terrestrial general interest programming creates the structure of the TV transmission market, composed of actors dealing only with the technical side of broadcasting. TV transmission operations were integrated into TV channels, and still are in some cases (in Italy for instance). The size of TV broadcasting companies is therefore now tied directly to the size of the TV markets as a whole.

The TV broadcasting sector has reached maturity on Europe's leading markets. The spectrum available for broadcasting TV signals is limited; digital compression and the switch-off analogue television will release additional spectrum in the long term, but without challenging the capacity provided by direct-to-home satellite or cable networks.

In addition, a series of factors is driving down prices:

- technical advances which make it possible to optimise broadcasting by reducing the number of sites, increasing the transmitters' power, and data compression rates; and
- growing competition, albeit still limited, comes from new entrants in terrestrial transmission, taking advantage of the launch of DTT.

The introduction of digital terrestrial television therefore constitutes a growth relay, at least during the simulcast period when both analogue and digital channels are being carried, as broadcasters will maintain two parallel terrestrial distribution networks.

DTT STRATEGIES IN EUROPE

The Spread of DTT

In terms of digital TV penetration, even in those countries which already have introduced digital television, the penetration rate remains fairly low. A maximum of 63 per cent has been reached in the UK, but all other figures are lower, demonstrating a very early take off stage. If we enlarge the picture and include not only digital terrestrial television but also digital cable and satellite, the forecast is of a steady trend of increased penetration in the next five years. It applies to the three platforms for diffusing digital television, but satellite has an edge over the two others and globally will reach 60 per cent of households by 2010 (IDATE, 2004).

Looking at Tables 4.1 and 4.2 we see a diverse picture, of a largely fragmented market.

The platform responsible for the introduction and ubiquity of television in most European countries, terrestrial broadcasting, is the last network to be digitised. In addition, although a growing number of countries have introduced digital on their terrestrial network, with the exception of the UK, DTT's rate of penetration among TV households is still very low.

Digitising the terrestrial network is proving a delicate matter, particularly in those countries where terrestrial reception is still the most popular mode, since it means that the majority of the population will be required to upgrade their equipment if they want to continue to receive their TV programmes. The transition from analogue to digital appears to be more demanding on viewers than the earlier transition from black and white to colour TV. This is, because the installed base of TV sets was not as big then as it is now, and because the programmes that began to be broadcast in

Table 4.1   DTT progress in Western Europe

	Legislation in place	Soft launch	Full launch	Switch off date
Operational platforms				
UK (October 2002 relaunch)	July 1996	September 1998	November 1998	2006 to 2012
Sweden	May 1997	April 1999	September 1999	2008
Spain (2004 relaunch)	October 1998	May 2000	May 2000	2012
Finland	May 1996	August 2001	October 2002	2007
Germany, Berlin	Spring 2002	November 2002	1Q 2003	ongoing 2010
Netherlands	1999	April 2003	4Q 2003	start 2004
Italy	2001	December 2003	2Q 2004	2006
Next to launch				
Germany (other regions)	2002	2004	2004	2010
France	August 2000	March 2005	2006	2010
Switzerland	2003	2005	2006	2015
Austria	2001	2005	2006	2012
Norway	March 2002	2005	2006	2006–2008
Portugal	2000			2010
Denmark	December 2002			2011
Belgium	2002			Flanders 2005
Ireland	March 2001			2010
Greece				

Source: EBU Report September 2004.

Table 4.2   DTT households and penetration

	1998	1999	2000	2001	2002	2003 1Q	2004 1Q	Penetration %
UK	247	552	774	1217	1260	1400	3000	12.0
Sweden		15	35	83	100	140	200	4.8
Spain			3	150	150	130	130	1.0
Finland				5	10	97	300	13.1
Germany						120	170	0.5
Netherlands						3	12	0.2
Italy							25	0.1
Total	247	567	812	1455	1520	1890	3837	

Source: EBU Report September 2004

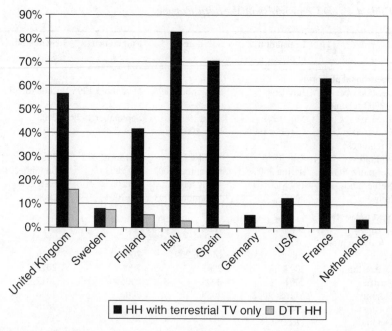

Note: HH: households.

Source: IDATE.

*Figure 4.2   Penetration of terrestrial TV, including DTT, in TV
households, in mid-2004*

colour could still be viewed in black and white on an older set. The migration therefore took place gradually, at the rate at which viewers replaced their sets.

Although digital terrestrial programmes can be received by terrestrial antenna, in theory without having to replace the antenna in most cases, current TV sets cannot receive digital programmes as they are transmitted. To receive DTT programmes, viewers therefore need to equip themselves either with a digital-analogue demodulator, which is connected to both the antenna and the set, or with an integrated digital TV set. The migration to digital terrestrial is therefore not as simple for viewers as was the migration to colour TV.

For consumers, the added cost involved in migrating to digital is not a major factor in most Western countries. The current price of decoder/demodulators (50–80 euros for a basic set-top box with no added features) makes them accessible to the vast majority of households, although an additional box is needed for each set or video recorder (VCR). The problem lies more in the rate at which viewers acquire this new equipment. Until the entire population is capable of receiving their TV programmes through a mode other than analogue terrestrial (that is via DTT, cable, satellite or ADSL), it would appear impossible to switch off the analogue signal. This means that there has to be a transitional period during which analogue and digital are simulcast, which constitutes a sizeable cost for all the channels involved (chiefly the so-called incumbent channels). It also means that additional frequencies need to be made available on the terrestrial spectrum during this transition period. Some countries are therefore able to deploy digital terrestrial only on a limited number of multiplexes, or in only a portion of the country (leading to problems of co-ordination at the borders). It is therefore in the interest of the channels involved in this dual broadcasting, in the state's interest, and also in the interest of the other users of the terrestrial spectrum, to make this simulcast period as short as possible. The challenge then is to provide an incentive for households equipped to receive their TV programmes over the analogue terrestrial network to migrate as quickly as possible to another reception mode. At the regular rate of equipment renewal, it would take 20 years for virtually all of a country's households to be equipped to receive digital terrestrial programmes, including multiple sets.

There are two options for speeding up households' migration:

- by leaving them no choice; or
- by providing them with an incentive.

The first case involves forcing TV viewers to migrate to another reception mode, or be deprived of television altogether. Here a deadline for the analogue switch-off would be set, after which analogue terrestrial programmes would no longer be available. It would therefore be up to viewers to act before the deadline, equipping themselves with digital terrestrial reception gear, or opting for another platform. This solution could be applied nationally (switch-off of the analogue signal nationwide on a given day) or locally (a zone by zone switch-off of analogue). In Italy, for instance, 31 December 2006 has been set as the deadline for the complete switch-off nationwide; in Finland, it is 31 August 2007. For its part Germany has opted for a gradual deployment of DTT, region by region. The Berlin-Brandenburg region thus became the first German region where the analogue terrestrial signal was stopped after only 9 months of simulcasting in 2004. It should be pointed out that, as yet, there is no single country where analogue broadcasting has been completely phased out. In cases where a sizeable portion of the population is still watching analogue terrestrial TV only, as the deadline is nearing, will the government go through with the switch-off anyway? In the US, even though the FCC had planned to recover the frequencies currently used by analogue broadcasting in 2006 in order to re-allocate them to other uses such as telecoms and emergency services, this timetable will not be met. And this is so despite the fact that analogue terrestrial broadcasting involves just over 10 per cent of the population. What will happen then in countries like France, Spain or Italy, where over 60 per cent of the population is supplied by analogue terrestrial broadcasting?

The incentive-based option involves encouraging viewers to switch to a service other than analogue terrestrial, by offering them more appealing programming – either in terms of diversity, quality or ease of access – on another platform, or by combining the analogue switch-off with certain financial incentives (for example reduction in TV taxes or subsidies for the purchase of digital equipment). But this option is hard to implement. Private operators provide satellite, cable and ADSL-based services. Although regulated by independent national authorities, these services cannot evolve based on government rulings. It is in the interest of these alternative operators to offer terrestrial viewers a service that is sufficiently appealing to make them want to switch. But while the logic of competition can help accelerate the natural migration from analogue terrestrial, it is not in itself enough to guarantee the migration of all analogue terrestrial households. This is particularly true since, in some places, it is virtually impossible to access an alternative platform (cable and ADSL are rarely available in rural or sparsely populated areas, and satellite is less popular with city dwellers). Added to this, very few alternative offers are free:

- viewers either need to take out a (usually monthly) subscription to receive their TV programmes;
- or the TV service itself is free, but to access it viewers have to subscribe to another service (such as broadband internet on cable or ADSL);
- or the offer is entirely free, but requires users to buy relatively expensive reception gear (for example a dish for free to air satellite channels).

Even though television has become a near ubiquitous form of mass entertainment, this does not mean that everyone is willing to pay for it.

For governments therefore, the only room to manoeuvre is with respect to the digital terrestrial service, and the possible advantages it can offer TV households. Terrestrial transmission is in fact the only distribution mode over which governments maintain a degree of control. They are therefore in a position to rule on those elements that could render DTT more appealing than the analogue terrestrial service. The US and Japan elected to impose HDTV (high definition TV) on the digital terrestrial network, thereby improving the quality of the service. For their part, Germany and the Netherlands have sought to promote DTT's portability, and even mobility, which constitutes an advantage in those countries where cable is extremely popular – allowing viewers to do away with cable and plugs, and move their sets freely from room to room.

The business models that governments choose will be equally decisive. Whereas most of those countries that have pioneered DTT had originally opted for a pay-TV model at launch, it soon became clear that it was not a viable option. Cases in point here are the failure of Britain's ITV Digital and Spain's Quiero TV services in 2002.

Other countries such as Germany and Italy set up systems to subsidise the purchase of DTT set-top boxes. In Germany, the support funds, fed largely by public and commercial channels being broadcast on the digital terrestrial network and by regional regulators, helped low-income households to buy a set-top box. In Italy, the government's 150-euro subsidy is awarded to any household wanting to buy a digital terrestrial STB, regardless of their income. Thanks to this programme, and to pay-per-view football, 650 000 set-top boxes were sold in Italy in the first 9 months of 2004. Regardless of the success of these measures, they are nevertheless questionable since they clearly discriminate against other broadcasting platforms.

Of course, there exists no single solution since each country's TV industry is different, as are each country's goals in this area. But, we can still distinguish two main types here:

- countries where terrestrial is the leading broadcasting mode, such as France, the UK, Italy, Spain and Greece; and
- countries where terrestrial broadcasting is secondary, or even marginal, such as Germany, Switzerland and the countries of Benelux and Scandinavia.

In countries where terrestrial broadcasting is still the most popular, the challenge is a major one since it involves migrating the majority of households in as short a time as possible. In addition to the huge number of households that need to make the transition, there is also the question of the digital terrestrial network's coverage. As it stands in these countries, the analogue terrestrial network – which was generally deployed using public money, at a time when alternative platforms either did not exist or were too costly to deploy – covered the entire population. But no country, apart from the United Kingdom, currently plans to have the entire population covered by the digital terrestrial network. When the analogue signal is switched off, what solution will be offered to the 10 or 15 per cent of the population that will never have access to DTT?

In countries where terrestrial broadcasting is secondary, the challenge may at first glance appear more manageable, since a smaller number of households are involved. But no country, even among those with the densest cable networks, such as Benelux, has done away with terrestrial transmission entirely. Migrating the terrestrial network to digital is even expected to help rekindle interest in it. While in countries in the first category, DTT is not a direct competitor for cable and satellite – particularly since it offers a much smaller selection of channels – in countries where so-called 'free' cable is very popular, DTT could prove a real threat. In these countries, TV households already enjoy access to some thirty analogue channels via cable, for a small monthly fee (around 10 euros). Should DTT offer a comparable service at roughly the same price (the Netherlands), or for less (Germany), and in digital quality, it could emerge as a serious rival. A mass exodus of cable subscribers to DTT does not seem very likely, however, and DTT in these countries is aimed chiefly at households with a second set, and those desiring portability.

Competition between DTT and Other Networks

As it stands, terrestrial television still has only limited capacity. In most European countries, the terrestrial network is capable of carrying only five or six analogue channels nationally. Use of the terrestrial spectrum is in fact limited by several factors:

- Not all of the available frequency bands can be used for TV broad-casting, either because some bands are not suited to it or because, in the portion that could be used for TV broadcasting, some of the frequencies are being shared with other applications.
- Use of the terrestrial spectrum also involves very close co-ordination with neighbouring countries. This means that some frequencies cannot be used as they could cause interference with a programme being broadcast from a bordering nation.
- Added to this, two adjacent channels cannot be used for ana-logue services in the same zone, since there is a danger here too of interference.
- Two neighbouring transmission sites cannot use the same frequency, even if it is to broadcast the same programme because, once again, there is a danger of interference in the zone where the two transmit-ters' coverage overlaps.

Nevertheless, in addition to channels that enjoy national coverage, it is possible to allocate frequencies for local programming, zone by zone.

The advent of digital makes it possible to alleviate some of these restrictions:

- Thanks to digital compression, it is possible to broadcast several channels on a same frequency. Generally speaking, depending on the rate of compression used, the expected quality and the bit rates required (which vary depending on the nature of the channel), between four and six standard definition channels can be broadcast on a multiplex in MPEG-2.
- The dangers of interference between adjacent channels do not exist in digital, which means that management of the terrestrial spectrum can be optimised by using two adjacent frequencies in the same zone whereas, on the analogue spectrum, gaps have to be created between them.
- In the same vein, it is also possible to use the same frequency on two neighbouring transmission sites, without any risks of interference.

The switch-off of analogue terrestrial broadcasting can only take place gradually in the majority of countries, and a long period of analogue–digital simulcasting will be required. This means that new frequencies need to be allocated nationwide to enable the parallel broadcast of both analogue and digital formats of TV programmes. Most European countries have therefore allocated between four and six national multiplexes for broadcasting DTT programmes, giving the possibility of broadcasting between 16 and 36 digital channels in standard definition.

Although there is a significant increase in the number of channels that can be broadcast on the terrestrial network thanks to digitisation, terrestrial capacity are still well below those of cable and satellite, with little prospect for improvement.

The introduction of more powerful compression systems like MPEG-4 makes it possible to free up additional capacities, either to broadcast new channels or new services. But, everywhere where DTT has been launched, the MPEG-2 standard has been used. Without even addressing the investments that a broadcaster needs to make to migrate from MPEG-2 to MPEG-4, the transition itself would require that the entire base of MPEG-2 set-top boxes in viewers' homes to be replaced, since they are incapable of receiving MPEG-4 signals. It appears impossible to undertake broadcasting in both MPEG-2 and MPEG-4 at the same time, not only because of the cost, but also because of a lack of available frequencies, so it will not be until the installed base of set-top boxes are MPEG-4-compatible that broadcasting in MPEG-2 can be phased out.

Furthermore, it would be possible to increase terrestrial's capacity by freeing up new frequencies. But, since the terrestrial spectrum is a scarce resource that is shared by a great many applications, it seems unlikely that new national multiplexes will become available before analogue broadcasting is switched off. A few countries have already set the deadline for the end of analogue, namely Sweden, Finland and Italy, but aside from the government's commitment to phasing out analogue, nothing guarantees that these deadlines will be met. In the United States, for instance, the FCC had set a proactive timetable of putting an end to analogue broadcasting by 2006, but we already know that it will not happen.

Any increase in terrestrial capacity arising from the use of more powerful compression and from new frequencies will not necessarily lead to an increase in the number of channels on offer. In fact, should HDTV become an integral part of DTT, it will assimilate any leftover capacity. DTT was launched in the US, Japan, Australia and South Korea, with obligations on the channels to broadcast for part of the day in high definition. In all of these countries, viewers capable of receiving DTT therefore enjoy HD simulcasts of the analogue terrestrial channels, but do not receive a broad selection of channels as they do in Europe. In all European countries, quantity won out, whereas the countries cited above chose quality over quantity.

A comparison of network efficiency should take into account the other services that these networks can distribute. Here, we suppose that only cable and ADSL are capable of making available marketing true triple play offers (TV, internet access, telephone).

As an illustration, we assume that the ARPU generated by internet access and telephony services could cover 50 per cent of the network's costs. Table 4.3 and Figure 4.3 show how the average cost of supplying a TV

Table 4.3   How the cost per channel of delivering service to subscribers depends on the level of penetration (€)

Yearly cost per channel and per subscriber (€)	Service penetration			
	10%	25%	50%	75%
ADSL	0.9	0.9	0.9	0.9
Satellite	1.0	0.7	0.6	0.6
Cable	1.0	0.5	0.4	0.3
DTT	5.2	2.8	2.0	1.7

Source: IDATE.

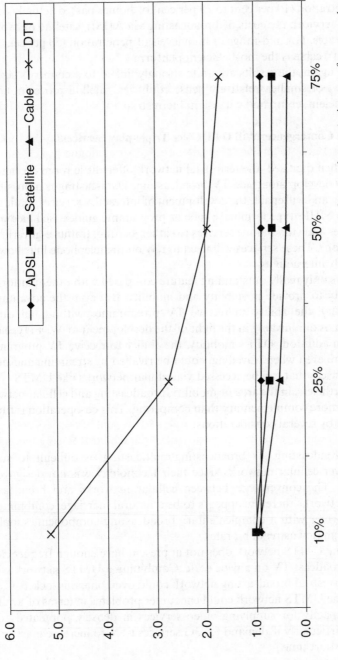

Figure 4.3 Comparison of the annual cost per channel per subscriber (€) according to the service's rate of penetration, by broadcasting mode (including other services)[2]

channel varies with the proportion of households served. For very low rates of penetration (10 per cent to 25 per cent of homes passed actually accessing the service), the costs of broadcasting via ADSL, satellite or cable are comparable. For mid-range or high levels of penetration (50 per cent to 75 per cent), cable is the most efficient platform.

Here again, the results are quite strongly linked to a country's size and density. For smaller, relatively dense countries, satellite proves not to be very efficient, compared with digital terrestrial TV.

DTT in Convergence: Will DTT Offer Triple-play Services?

Even when digitised, the terrestrial network offers little room for new services to develop alongside TV broadcasting. Data-casting is a possibility, however, and will enable the development of interactive services, made available via MHP[3] type terminals, such as programme guides, rich media programmes and stand-alone services (weather, betting, traffic, e-government, and so on). These services will need to rely on the telephone line, however, to supply interactivity.

DTT's only really outstanding feature compared with other platforms is its ability to provide portability and mobility. But even the advantage of portability, the ability to receive TV programmes without an outside antenna, is diminishing in the light of the development of Wi-Fi type solutions. In addition, DTT's mobility, the ability to receive TV programmes anywhere even when travelling, could be rivalled by streaming and download services that can be accessed via cellular networks like UMTS.

Nevertheless, in the area of mobility, broadcasting and cellular networks appear more complementary than competing. This co-operation is further boosted by several considerations:

- Broadcasting, or broadcasting/multicasting, is difficult to supply over cellular networks since their technology was not designed for it. The convergence between cellular networks and broadcasting networks therefore appears to be a natural marriage, combining the former with a complementary broadcasting component, supplying high downstream bit rates.
- The UMTS network does not at present have enough frequencies to broadcast TV on a wide scale. Combining a UMTS network with a terrestrial broadcasting network could overcome this lack.
- The UMTS network could encounter problems in terms of available capacity for supplying video services in densely populated zones, particularly if demand for 3G services were to increase massively in a short time.

Distribution in broadcast mode allows users to receive full-length programmes for free (excluding pay-TV channels) on their mobile terminals,
unlike distribution via streaming or download. On cellular networks, the
bulk of the cost of content downloads is not the cost of the content itself,
but the cost of data transfer. Even if subscribers know the cost of the programme they want to watch ahead of time, they will have no control over
the download cost, which is proportionate to the number of packets of data
it contains. In live streaming, currently available services are often based on
time-based billing systems (with flat rates for a given number of hours).

In the longer term, and particularly with the end of analogue broadcasting, it would be possible to use a portion of the available frequencies to
develop push offers, using a personal video recorder (PVR). The principle
here would be to take advantage of the added capacities to download a
selection of films or programmes to the hard drive, which viewers could
then watch when they want.

The Prospects for DVB-H

The idea has arisen of using DVB technology for broadcasting television
to mobile, handheld devices (the H in DVB-H meaning handheld). It is
therefore geared to equipment such as PDAs and mobile phones, enabling
a 3G mobile to receive video content in point-to-point broadcasting mode.

One of the advantages of this type of solution is the ability to use hardware developed for digital terrestrial TV, which makes it possible to save
on development costs and to benefit from the low prices associated with
DVB-T production prices (the DTT standard). In addition, even though
DVB-T was not designed for mobile use, trials have produced very positive
results, which have led to the launch of services notably in Germany and
Singapore, and to a great many commercial trials.

An additional standard was required (instead of just using DVB-T)
because of problems with the devices' battery when associated with a reception tuner. The DVB-T specification would drain the batteries very quickly
if used for TV reception over a mobile phone. This specification could also
require two antennae for mobile reception.

Other constraints were imposed on DVB-H: the capacity to receive
15 Mbps on an 8 MHz channel, and operation in SFN (Single Frequency
Network) mode, while maintaining maximum compatibility with DVB-T.
Lastly DVB-H can coexist with DVB-T in the same multiplex. This means
that both DVB-H and DVB-T services can operate on the same multiplex.

Several pilots have taken place in Europe since late 2002, and others are
planned for 2005–2006. The first project, Finland's MobTV was launched
in late 2002, with the chief goal of researching, developing and testing the

use of and applications for mobile TV, bringing together players from the world of research (VTT Information Technology, University of Tampere), a telecommunication company (Telia Sonera) and players from the media sector (Alma Media, Sanoma WSOY). The project did not have any stated short-term commercial goals. Finnish company RTT Oy conducted a technical trial in Helsinki in late 2003, which involved the distribution of internet content and digital TV via the DVB-T system (using IP Datacast technology), and which was expected to evolve towards DVB-H. Finland's communications regulator awarded this project a three-year licence for testing the network. Another Scandinavian project, called Finpilot, was launched in late 2003, led by French company TDF Group in the city of Helsinki.

Since late April 2004, as part of the European R&D project INSTINCT, TDF has also been running broadcast tests on mobile TV in DVB-H mode in the French city of Metz. Through this pioneer experiment in France, in which several international partners are taking part, TDF is involved in fine-tuning the technologies that will enable the broadcast of real TV channels on mobile devices.

After announcing their partnership in August 2003, Nokia, Philips, Universal Studios Networks Deutschland[4] and Vodafone's research centre in Germany announced the BMCO (Broadband Mobile Convergence) project at the CEBIT show in Hanover in March 2004. The goal of the project is to create a convergent content platform geared to mobile devices. The platform combines GPRS (2.5G) type mobile communications and digital terrestrial broadcasting. This project launched in late 2004 in Berlin. It aims to demonstrate a model for co-operation between UMTS and DTT networks, to standardise DVB-H technology and to establish a viable business model. The platform that will ensure intermediation between these two networks and service billing is not expected to become a reality in the next five years.

The latest European project, the Dutch digital terrestrial TV platform, Digitenne, in partnership with Nokia and research institute TNO, broadcasts video content to mobile handsets. The trial phase was in May 2004, in the western part of the country. Marketing of the service is not expected until 2007 at the earliest.

THE IMPACT OF DTT

In addition, what we have at the moment is a fragmented and widely diverse DTT market in the European Union, although it is supposed to achieve a unified internal market. Plus there are many challenges for digital terrestrial television in Europe. In the first months of 2005, however, (Business Week, 2005) some success has been achieved, especially when compared with the US.

One of the incentives and one of the rationales for the transition from analogue to digital television was expected savings in radio spectrum. In terms of economic calculation at the firm level, however, there is no real incentive, as most television broadcasters do not pay for the radio spectrum they use. Most debates on this issue enter a vicious circle. When people say to television broadcasters: you should save on the spectrum, or maybe you should even pay for your spectrum, what you get is an uproar of contestation by television broadcasters saying, if we have to pay for our spectrum, then we should be relieved of our public service obligations, like unbiased news, full geographic coverage, and so on. A linkage has been established between broadcasters not paying for the spectrum and their being subject to public service obligations. And (for the moment) no escape from this vicious circle has been found. A regulatory bias has been introduced.

This compounds with the lack of a clearly defined business case for DTT. The overall amount of household expenditure available, a combination of public funding, advertising and subscription and pay TV revenues, is stable, or increasing only very slowly (with the exception of countries where a catch-up process is underway). This limitation puts TV broadcasters on a collision course with the glut of competing technologies able to provide similar services. This variety of technologies is at the very core of the rationale for convergence, but in this case, the TV industry has a weak incentive to add a new platform to analogue, plus satellite, cable and the advent of TV on ADSL. To give an example, the biggest television broadcaster in Europe at the moment, the France–based TF1, says it does not want to go to digital terrestrial television: it has enough. Broadcasters have enough to do, dealing with cable, which is struggling, with satellite, which is expanding but needs very high marketing expenditure, and with the possibility of using ADSL, which appears to them to be cheaper than DTT. Europe does have converging and competing technologies, but we have too many of them from the viewpoint of TV channels.

Another issue to be mentioned here in passing is the difficulty of defining fair treatment between public television and private television. The UK provides an interesting experience. It has been flexible enough to retreat from an initial failure of the DTT pay-TV business model. But apparently, perhaps because of the different contexts, other European countries have not learnt the lesson, Each European country is apparently coming up with its own recipe for the financing of the transition to DTT. Those recipes are very diverse, combining public and private initiatives, incumbent television channels and new entrants, free-to-air and pay TV channels, and so on. The recipe is very complex in each country and there is no obvious key success factor.

Does this build a case for strong public intervention? It is not clear either. Assume that there is public intervention in one domain, for instance in favour of public channels, or just because spectrum is free for television. The problem is, that once that bias in market mechanisms is introduced, it takes hold and generates other biases, and it becomes very difficult to handle them all. This interacts with the many dominant positions of firms in the various segments of those converging markets. From a corporate strategy viewpoint as well, there are many innovative dilemmas – that is situations where a new technology would obviously be superior to the current one, but introducing it would kill the existing business. Although competitors are coming, or are already here, companies find it hard deliberately to sacrifice their existing business.

Whatever those difficulties and challenges, there are very good grounds for public intervention in the name of technical platform neutrality. This would be public intervention to eliminate public intervention. Previously Europe had different regulatory frameworks in the various areas on cable, television, telecommunications, and so on. In 2002 the European institution enforced a common regulatory regime for all electronic communications services, excluding content (EC, 2002b). This new regulatory framework came as a welcome relief to regulators, because the previous prevailing concept, facilities based competition, has only been partially successful. We must remember that, in the first stage of telecommunications markets opening in the 1990s there was a debate about service-based versus facilities-based competition. Most regulators thought that facilities-based competition was preferable and might be more practicable, but it just did not happen. With the very notable exception of mobile telephony, there has not been widespread duplication of similar networks. This was at odds with what most regulators were contemplating in the 1990s. The convergence of technologies has been quite a welcome phenomenon, as it allows the possibility of inter-platform competition. Inter-platform competition permits competition even where strong elements of monopoly are still prevalent in individual pre-convergence platforms.

The difficulties faced in implementing technologically neutral regulation in inter-platform competition are still considerable. Just to give an example of the distance the regime must travel to come up with a consistent regulatory framework, consider the current difficulties in the application of the must-carry rule, and its distinction with open access. The latter is a notion applicable to some technologies, some networks. Not all platforms should provide open access. For instance in the case of television, the goal of neutrality among the various technologies has confronted regulators with problems: the open access principle prevailing in telecommunications networks is not yet fully part of television regulation. There is no general open access obligation. Operators are only required to convey programmes of particular importance to society, and which are provided for the benefit of all consumers. For instance, satellite TV operators in France have not been obliged to convey all new DTT channels, thus strongly restricting their viewers at an early stage.

Another important issue is inter-operability (EC, 2005b). In practice, the interoperability of digital interactive television services has not yet been achieved in Europe. For example, consumer equipment made in one country cannot be used to receive services in other countries. Within some countries, furthermore, there is a plurality of proprietary platforms, and so on, which hinder the citizen's reception of competitive services. The lack of interoperability has been an obstacle to the development of pan-European markets. Today, in technological terms, we have 'television with frontiers'.

The objective must be for the citizen to have access to the maximum range of content. This means that open technical standards are preferable and that the emergence of digital gateways should not hinder universal access to a common, shared sphere of media services.

Only interoperability can fulfil the objectives explicitly mentioned in Article 18 of the Framework Directive – to facilitate the free flow of information and ensure pluralism of media and cultural diversity.

At the European Union level, this notion is very central. Whether in relation to computers, software or networks, it is asserted that inter-operability is necessary for competition to be effective. This is very crucial, but it has not yet been accomplished. Great efforts have yet to be made to actually achieve inter-operability within the various platforms (EC, 2003b).

DTT may also affect the regulation of terrestrial broadcasters. As a counterpart to the access to the scarce spectrum resource, public and private broadcasters faced public service obligations such as programming and production quotas. In most European countries other cable and satellite channels benefit from a lighter regulatory regime. As DTT enables the entry of more players into the terrestrial television market, including players already active on cable and satellite, maintaining a specific regime for terrestrial television may be harder to justify.

Given all these challenges and difficulties, the basic concept of convergence itself must be examined. This is a debate which occurs not only at the international level, but also within the European Union itself.

There is not one but two concepts of competition, based on how playing fields are defined. For some people the playing field should be level, as in a cricket game for instance, in the sense that there should be a uniform legal framework. But there is also a different notion of competitive playing field, which might be called the golf or Tuscan Hills playing field. According to this conception, competition occurs not on an individual market, but between the various market frameworks. Each country and each region offers to market players in the world economy, a certain combination of economic resources, social resources, infrastructures and a legal framework. There is then competition between those legal frameworks. If you do not like the legal framework in Hungary, for instance, then you move to China. In this particular conception, uniform rules are not required; competition is between the rules themselves.

In relation to television transmission there are multiple technologies. This is both good, because diversity and convergence promote competition, and bad, because companies might balk at using them all. There is also the question mark over the expansion of the market. At the high point of the 1990s bubble, the introduction of DTT was supported by the notion that market spending was growing. But if spending is not expanding, what does it mean for the future of DTT? There are also question marks over the extent of the future convergence of the various technologies, including television broadcast on mainly voice-oriented terminals, like telephones. Conversely, extended interactive functionalities on TV sets would manifest the realities of convergence.

How far these potentialities will be realised is still an open question. If convergence of services remains only a possibility with telephones remaining mostly telephones, televisions remaining mostly televisions, PCs remaining mostly PCs, what is the meaning of convergence in industrial terms or for regulators? What is its effect on competition? Convergence is still a very desirable outcome for citizens, for consumers, and for regulators and governments, as a means to the end of greater competition. It is obviously desirable, but that does not make it happen.

Finally, does this create a new momentum for free television? The difference between the number of free-to-air channels received by consumers is the most fundamental difference between Europe's different markets. The launch of digital terrestrial TV, together with television over ADSL, will contribute to shrinking this gap.

In countries where only a very few free channels are currently being offered, the swift conversion of households that receive fewer than 10 free channels to receiving more than 20 will be the major development on TV markets over the next five years. But they will not all survive. A period of concentration of the channels will be needed for the market to achieve overall profitability.

CONCLUSION

As a conclusion, if we look at the evolution of the television landscape over the last decade, it appears that in relation to the general audiovisual objectives related to content diversity, quality and universality, two concerns have recently been added or reinforced. Digital technologies have made the right level of intellectual property rights protection and digital rights management of critical importance, and the promotion of a competitive television market with open interoperable standards, especially in the context of competing digital platforms, has taken centre stage.

Those two issues will not be easy ones to address. Three screens are now competing for TV content: the regular TV set, the PC, and now the mobile handset. Conflicting interests between them abound, and inconsistencies between telecom and media regulation, like those encountered in spectrum assignment procedures, are not easily overcome.

NOTES

1. Case 155/73 *Giuseppe Sacchi* [1974] E.C.R. 409 [1974] 2 C.M.L.R. 177.
2. For this figure, half of the network costs of ADSL and Cable are supported by other services other than television (voice, internet access).
3. Stands for Multimedia Home Platform, A standard for Interactive television.
4. A subsidiary of Universal Studios in Germany.

REFERENCES

Business Week (2005) 'The TV revolution sweeping Europe', 1 April.
EC (1989) *'Television without Frontiers' directive* (89/552/EEC) amended on 30 June 1997 by the European Parliament and the Council Directive 97/36/EC.
EC (2002a) *Fourth Communication from the Commission* (COM (2002) 778 final).
EC (2002b) 'Directive (2002/21/EC) on a Common Regulatory Framework', *Official Journal*, OJ L 108, 24.4.2002, p. 33.
EC (2003a) 'Communication from the commission to the council, the European Parliament, the European Economic and Social Committee and the Committee of the Regions on the transition from analogue to digital broadcasting', *'From Digital 'Switchover' to Analogue 'Switch-off'* (SEC(2003)992).
EC (2003b) *Communication From The Commission to The European Parliament, the Council, the European Economic and Social Committee and the Committee of the Regions on Barriers to widespread access to new services and applications of the information society through open platforms in digital television and third generation mobile communications*, Brussels, 9.7.2003 COM(2003) 410 final.
EC (2005a) *eEurope 2005: An information society for all*, COM (2002) 263 final.
EC (2005b) *Contribution to public consultation on the challenges for the EU's Information Society policy beyond*, European Broadcasting Union, 14.1.2005.

IDATE (2003) *Digital Terrestrial Television Market Report.*
IDATE (2004) *TV in Europe 2015 – Consumer trends.*
Perrucci A. and A. Richeri (2003) *Il mercato televisivo italiano nel contesto europeo*, Collana 'Studi e Ricerche', 88-15-09367-2.
World Economic Forum (2005) *Global Information Technology Report 2004–2005*, 2003.

5. The development of digital broadcasting in Italy

Alfredo Del Monte

INTRODUCTION

Broadcasting regulation has been justified in many different ways: economically, culturally and politically. Spectrum scarcity, economies of scale and scope, or low elasticity of substitution with alternative services and products can lead to monopolisation. Other sources of market failures are asymmetric information and biases arising from advertiser finance, and the existence of externalities, both positive and negative. Without intervention the broadcasting market would fail to deliver the socially optimal mix of programmes.

While different countries have shared a common conviction of the need for broadcasting regulation, they have followed diverging policy approaches. Such approaches have been largely dictated by national and political administrative traditions and have resulted in different structures for the analogue TV sector. Some authors, including Galperin (2004), think that it is important to analyse the structural features of the TV analogue sector as they constrain the implementation of the transition strategy to digital:

> Individuals and organizations make long term commitments (i.e. investment in particular broadcast technologies or services based on the existing rules of the game). Because these commitments often represent sunk costs, these market agents tend to resist policies that significantly alter these rules. This facilitates policy choice consistent with existing regime and inhibits those deflecting from it (Galperin, 2004:18).

A shift from one regime to another is possible, but it requires mobilisation of large political resources and not many governments are able to resist the political pressure of interests linked to the current structure of the TV sector. Technology is only one of the factors that affects the structure of the TV industry and is not necessarily the most important. Therefore the transition to digital TV very often follows a pattern determined by the pre-existing structure of the terrestrial analogue sector. The transition to digital TV in Italy lends support to the above hypothesis.

This chapter has two aims. The first is to show how the current features of the Italian TV industry are the consequences of past regulatory intervention. The second purpose is to analyse how the current structure of Italian industry affects the policies to promote the switch from analogue to digital TV, and to consider whether these policies will be able to bring competition to the Italian TV market.

DIGITAL TERRESTRIAL TV AND THE STRUCTURE OF THE TV MARKET

Digital technology has revolutionised the telecommunications and the information industries, but until recently its impact on the television industry has been much weaker. However, there is a widespread belief that the same forces that in recent years have radically changed the telecommunications and information technology will do the same for broadcasting, where the pace of technological change has been much slower.

The transition to digital TV is much more advanced in satellite and in the cable TV industry than in analogue terrestrial TV. Hence the structure of the terrestrial industry will be transformed by the introduction and the development of digital terrestrial television (DTT). The main characteristics of DTT technology are as follows:

1. a much better movie picture quality and CD sound quality;
2. rapid delivery of a large amount of information and the possibility of receiving personalised programmes and interactive services;
3. DTT allows better use of broadcast spectrum potentially, increasing its availability to new broadcasters and providers of other information and communication services;
4. DTT technology allows a single programme to be broadcast at a lower cost and easier separation between network providers and content providers; and
5. DTT allows the spread of conditional access systems, enabling broadcasters to offer a large package of digital products on a pay-TV or pay-per-view basis.

The relaxation of spectrum scarcity should decrease barriers to entry and therefore increase competition in the television market. The higher number of channels should increase programme differentiation. Some authors (for example Noam, 1998) think that commercial broadcasters would supply high quality programmes provided the numbers of channels is large enough. A market-based system cannot be expected to supply additional public interest programmes, as they would not be profitable. There might be an incentive to supply some specialised categories with a high content of violence and pornography unlike conventional programming. For this reason regulation of digital TV still seems to be required.

Another important effect of DTT will be the reduced weight of the free-to-air model *vis-à-vis* pay-TV and pay-per-view models. The increase in the number of channels will reduce the average audience per channel and competition could reduce the price of TV advertising. Therefore we could expect profits of free-to-air broadcasters to decrease. Total expenditure by advertisers may well decrease as technology (for example personal video recorders) allows customers to record TV programmes and skip the advertising.

Another important factor that poses a threat to the free-to-air model built on the 30-second spot that fetches a premium at prime time is the changing policy of advertisers. Once an advertiser could reach most customers with an advertising spot aired simultaneously on the main free-on-air networks. Today this is much more difficult. In developed countries the mass market has atomised into market segments defined not only by demography, but by increasingly nuanced product preferences. The proliferation of digital and wireless communication channels allows advertising companies to reach specific market segments. It will be increasingly difficult for traditional networks to reach the same number of customers, when there is increasing audience segmentation among different platforms and channels.

The evolution from mass to micro-marketing will probably lead to a decline in the mass media's share of advertising. Online advertising could threaten the traditional advertising markets. The top portals could reach truly mass audiences; the Internet is expanding, and banner ads could reach a large number of specialised markets more easily than the old TV model. It is highly probable that the Internet will increase its share of media advertising expenditure, and commercial TV will decrease its share. Pay TV and pay-per-view are therefore possible alternative revenue streams for large TV networks. DTT with conditional access allows new methods of payment to see an event. Prepaid cards placed in the decoder could be used to see a single event (sport, films and so on) without the need to subscribe to specific channels.

In many markets the price mechanism plays the role of revealing information about consumer preferences. In the free to air model the absence of prices weakens the possibility of knowing consumer preferences precisely. Direct charging mechanisms (as in pay-TV and pay per view) allow more information about consumer valuations to be revealed and TV networks could broadcast programmes more tailored to consumer preferences.

DTT technology implementing pay-TV and pay per view decreases information asymmetry between broadcasters and customers.

DTT technology has features that could generate more efficient use of
spectrum, decreases in concentration, increases in pluralism and the introduction of new innovative services. The question is whether market forces
alone will bring such results.

To answer such questions we must examine two different aspects: concentration and programme differentiation, and pluralism and social objectives.
Concentration and programme differentiation are economic problems.
Pluralism and social objectives are political problems. Digital technology can
increase competition without increasing pluralism or achieving more important social objectives. Because television is pervasive and powerful and has a
large influence on economic and political preferences of its audience, public
interest groups have argued that the television industry must remain regulated.

This chapter focuses on the economic aspects of TV regulation. It analyses whether digital TV will decrease concentration and increase programme
differentiation with special reference to Italy. The transition strategy
adopted to switch from analogue to digital TV is very important both for
the success of the introduction of DTT, and for the strengthening of competition in the market. It will take a number of years to convert fully to
DTT because today's television sets are not designed to receive digital
signals. Therefore consumers must have an incentive to invest in new sets or
decoders. In order to provide DTT service, while continuing to broadcast
their analogue programmes, broadcasters may have to modify their transmission towers and construct new ones. Broadcasters must also invest to
produce (or purchase) and distribute more digital programming. Such
investments are necessary to encourage consumers to invest in upgrading
their receivers or buy decoders. The diffusion of DTT can be seen as a
chicken and egg problem. If the installed base of TV sets capable of receiving DTT signals is negligible, broadcasters have no incentives to invest to
implement DTT. On the other hand, in the absence of content, consumers
have no incentive to make investments to see DTT programmes. In this situation government intervention could be necessary to co-ordinate network
operators, programmers, equipment manufacturers, and to give incentives
to consumers to buy decoders or new receivers.

But there are other important policies that could affect market
concentration:

(a)   the allocation of licences between incumbents and new entrants;
(b)   the method used to allocate licences;
(c)   the obligations imposed on broadcasters that acquire or purchase the
      licences;

(d)   obligations to surrender analogue frequencies;
(e)   the conditions for access to distribution networks; and
(f)   the conditions for access to premium content.

Governments could choose strategies that have a different mix of the above policies. Such strategies will be the result of economic and political pressure. Incumbents in the analogue TV market can be expected to fight to delay the introduction of digital TV or to dominate the new digital market. In the transition period new DTT services and current analogue programmes will be available simultaneously. Therefore the spectrum used for analogue transmission will be unavailable, and additional broadcast spectrum will be needed to introduce the new DTT services. Only at the end of the transition period will incumbents be required to surrender frequencies used for analogue transmission and only then might spectrum be available for new entrants.

If the allocation of licences is not designed to tilt the market in favour of new entrants and restrain dominant operators, the incumbents will preserve and eventually strengthen their position in the new digital market compared with the analogue. The incumbent operators benefit from key assets such as distribution infrastructure (towers, frequencies, and so on), content (sport and film rights, and so on.); their image is based on an historical presence in the market which could strengthen their market position in the transition period (when there is still spectrum scarcity) and make entry more difficult when additional frequencies are available.

This is what probably will happen in the Italian case. Digitisation will not necessarily produce convergence between Italy and other nations as national industries adopt new technologies.[1] Such convergence cannot be taken for granted, and strong political and economic pressure is required for it to be achieved.

THE EVOLUTION OF REGULATION IN ITALY UP TO 1990

Italy is a country that until 1974 had no private or commercial television. Commercial television started in 1974 when the Constitutional Court delivered a judgment that a public monopoly of TV was not constitutional. In 1976 the Court delivered another judgment: public monopoly of TV was constitutional at national level but local TV channels were free to enter the market.

The government was also invited to regulate the spectrum assignment and the number of licences, but it never intervened, and spectrum allocation was determined by private broadcasters. As a result the number of local TV channels greatly increased, and in 1981 another judgment by the Constitutional Court opened the national market to private broadcasters. During the 1980s there was no regulatory intervention to allocate spectrum and licences. As a consequence, the structural evolution of the industry was left to the market. In the first half of the 1980s the private RTL SPA group (later called FININVEST) controlled three national television channels, and was the only private broadcaster able to compete with the RAI, the public broadcaster. In Italy in this period, the only interest of government in TV regulation was the political control of public broadcasting. Political pluralism in public broadcasting was regulated by a Parliamentary Commission which showed control with opposition political parties.

The results of this policy of the Italian Government were:

1. a very high concentration of the TV sector, with only one private broadcaster competing with the public broadcaster (Table 5.1);
2. a very high number of local TV channels;
3. inefficiency in spectrum allocation (for example interference problems); and
4. no pay-TV broadcaster.

The fairly large number of national and local channels,[2] the large number of free channels, the high costs of cable distribution, and the inefficiency of the local authorities in giving approval regarding digging, construction and so on made the development of pay-TV unviable (on both cable and satellite platforms). This constituted a great difference with other countries that by 1980 had already experienced rapid development of satellite and cable distribution.

ITALIAN BROADCASTING POLICY SINCE 1990

Italian broadcasting regulation changed at least formally after 1990. Until that year the market had been regulated by judgments of the Constitutional Court, without any active regulation by the Italian government. The Constitutional Court describes the situation of the Italian television market in this period as like the 'Wild West'.

In 1990 Law 223/90 was approved. Its intent was to regulate the Italian broadcasting market. It established media ownership limits that did not differ from those exhibited by the market structure in 1990. The number of national channels any company could own was three.

Table 5.1 Market shares in the terrestrial analogue broadcasting sector

Broadcasters	Towers and frequencies[c] (%)		Population coverage 2002		Audience (%)		Advertising revenues (%)	
	1990	2002	Potential	Actual	1990	2002	1990	2002
RAI	38.9	45.8	99.1	96.6	50.5	44.9	29.7	27.69
MEDIASET	41.3	35.5	97.7	71.6	35.9	42.6	60.9	63.01
Telecom Italia (Cecchi Gori)	7.25	8.3	90.8	47.8	1.1	2.2	3.1	3.03
Holland Coordinator – TF1 (Tele+)	6.75	5.1	66.6	33.9	} 12.5	} 10.3	} 6.30	} 6.27
Fondo Convergenza	2.64	2.1	83.5	28				
Gruppo Peruzzo Editore	1.06	1.3	80.3	34.3				
Telemarket	1.00	1.4	63.8	22				
TBS	1.10	0.5	51.1	13.5				
Total national broadcasters	100	100			100	100	100	100
Herfindhal index[a]	0.333	0.346			0.384[b]	0.386[b]	0.46[b]	0.475[b]

Notes:

[a] A Herfindhal index measures concentration on a scale from 0 (very large number of firms, each with a very small share) to 1 (a single monopoly firm only).

[b] Audience and advertising Herfindhal indices are based on share of the first three networks as we have no specific data for others.

[c] The first two columns measure the share of frequencies and towers used by each broadcaster as a percentage of total available spectrum and towers.

Source: AGCM and author's calculation

The law also fixed the total number of licences at national level (12). These limits allowed the main private (MEDIASET) and the public (RAI) broadcaster, to keep the number of channels they already had. There was no audience share rule, frequency share rule or advertising share rule. Therefore in 1990, RAI and MEDIASET shared the Italian broadcasting market. RAI, the public service broadcaster, had 38.9 per cent of the frequencies, 50.5 per cent of the audience and 29.7 per cent of advertising revenues: MEDIASET had 45.8 per cent of the frequencies, 35.9 per cent of the audience, and 60.9 per cent of advertising resources. The Herfindhal index for the Italian market was 0.333 in the case of frequencies, 0.384 for audiences, and 0.46 for the advertising market. The most peculiar aspect of the Italian broadcasting market is the very high concentration in the frequencies market. In a market typified by a limited spectrum, such as analogue broadcasting, a high concentration of frequencies creates high barriers to entry and reduces competition. This high rate of concentration in the frequencies was the result of the 1975–1990 period when frequency assignments was not regulated.[3] Furthermore, as the Autorità Garante per la Concorrenza ed il Mercato (AGCM),the Italian Antitrust Commission, has found (2004: 21), RAI had an excess of frequencies relative to its needs and probably the same is true for MEDIASET. This suggests that RAI and MEDIASET strategically bought frequencies in order to reduce the number of future entrants at a national level. Table 5.1 shows that RAI and MEDIASET competitors had, in 1990, only 19.8 per cent of total frequencies and probably an effective coverage of the total population of less than 40.5 per cent.

The high concentration of frequencies has led to a high concentration of audiences and of advertising revenues. Broadcasters sell their audience and part of their broadcasting time to advertisers. The larger the audience of a particular TV channel, the more attractive the channel is for advertisers, and the greater their willingness to pay for having advertising spots inserted in the programme mix selected by that channel. Table 5.1 shows that this strong correlation is found in the Italian television market, but that there is no direct proportionality between the audience and the advertising share. The public service broadcaster, RAI, had a much lower share of advertising than its share of audience, since there are more restrictions on advertising imposed on a public broadcaster than on private ones.[4] On the other hand the very small private companies, serving local markets, had lower advertising market shares than their audience shares, since they were not attractive to the large advertising companies. In contrast MEDIASET and the other, quite small, national broadcaster Cecchi Gori, had a higher share of the advertising market in 1990 than of the audience market.

A new law, 240/97, was approved in 1997, under the centre-left government. Its purpose was to increase competition in the Italian television market. Law 240 reduced ownership limits imposed on a single company. The main effect of the law, given the availability of 12 channels, was to reduce from three to two the maximum number of channels that a single company could broadcast. The private broadcaster MEDIASET was affected since it owned three channels.

Law 240 fixed the ownership threshold at a share of 30 per cent of the advertising market for a given sector (television, radio, newspaper industry) or 20 per cent of total advertising revenues of television, radio and newspapers combined if there were media's cross-ownership.

Law 240 was the basis of regulation of the broadcasting industry made by the Italian Communications Regulatory Authority (AGCOM) until 2004 (including Resolution 78 of 1998 on terrestrial analogue licences, and Resolution 435 of 2001 on digital terrestrial broadcasting). However it was never really implemented for judicial and political reasons; Silvio Berlusconi, the head of the opposition till 2001, and subsequently head of the Italian government, was the owner of MEDIASET.

Thus if we compare the structure of analogue television industry in 1990 with that in 2002 we see no decrease in the rate of concentration. There was in fact an increase in the Herfindhal index for frequencies, audience and advertising. The largest increase (from 0.46 to 0.475) was in the advertising market whose index increased. The public broadcaster RAI, the second private broadcaster Telecom, and local broadcasters had falling shares of the advertising market; only the private broadcaster MEDIASET increased its share. Probably this was due to MEDIASET's good performance in gaining audiences: its share increased from 35.9 per cent in 1990 to 42.6 per cent in 2002. In Italy unlike other European countries, the weight in the terrestrial analogue market of the two main broadcasters, RAI and MEDIASET, is not challenged by other platforms. Cable TV is represented by a small company FASTWEB, with less than 200 000 subscribers. There is stiffer competition for audiences coming from the digital satellite platform. The significance of this platform (see the following section) has increased in recent years but it is still too small to challenge the oligopoly in the terrestrial analogue market.

PROBLEMS AND OBSTACLES TO THE DEVELOPMENT OF THE PAY-TV MARKET IN ITALY

The pay-TV market in Italy started in 1994, when Telepiù, belonging to the French group Canal Plus, began to broadcast analogue satellite pay-TV. Development of pay-TV has been highly dependent on access to live rights to soccer and other sports. Initially subscribers could see a 'live' football match on Sunday evening. In 1997 Telepiù went digital. Thus it had the ability to broadcast soccer matches simultaneously. Live sports rights became the key to the success of digital pay-TV, as only subscribers could see key sport events. Since Telepiù could pass its programming costs on to a self-selected audience, it was able to make a higher bid than free-to-air broadcasters for sports rights, thereby increasing the incentives for dedicated fans to subscribe to its services.[5] In 1998 Stream, a company created in 1993 by Stet SPA, a state-run company, to offer interactive media services (video on demand, pay per view, and so on), entered the digital pay-TV market. Until 1997 Stream had been controlled by Telecom Italia owned by Stet; and Laws 223/90 and 224/97 did not allow a public sector company, other than RAI, to broadcast its own programmes. Therefore Stream was a service provider that carried on cable the signal for channels and programmes produced by third parties. In 1998 Telecom was privatised and therefore Stream, initially a cable company, had the possibility to offer a pay-TV service, or both with satellite and cable platforms. In 1999 News Corporation became a minority shareholder (17 per cent) of Stream.

The entry of Stream on to the pay-TV market led to a bidding war with Telepiù for sports rights. In order to protect the weaker company (Stream) from being excluded from broadcasting the main soccer events, the regulator restricted to 60 per cent the share of exclusive live rights to *Serie A* (Premier League) soccer matches that a single company could acquire (Law 78/1999). Therefore Telepiù and Stream shared the sport-rights market, Telepiù having the exclusive live rights of northern Italian clubs (whose audience was distributed throughout Italy) – Juventus, Milan and Inter – and Stream having the rights on popular clubs, though with fewer fans, such as Parma, Fiorentina, Roma and Lazio. The competition for football rights increased their prices steadily. The estimated prices paid for football rights in Italy rose five fold between 1998/1999 and 1999/2000.[6] After that period the increase was much lower (8.5 per cent between 1999/2000 and 2000/2001). In this period we do not find a similar increase in the number of subscribers and revenues of pay-TV companies.

The growth of subscribers accelerated from 20 per cent in 1997 to 29 per cent in 1998, and to 45 per cent in 1999. In 1999 the number of subscribers was 1.7 million (Table 5.2). The reason that the number of subscribers was much lower than in other countries were competition in a free-to-air sector that offered a large choice of channels, even if the quality was not very high, and the widespread diffusion of pirated cards for decoders.

On the other hand, competition between Stream and Telepiù did not allow a substantial increase in the subscription price. Hence the increase in costs of sports rights had a negative impact on the profits of pay-TV companies. In 2000 Stream had an average revenue per subscriber of 158.3 euros; the cost

Table 5.2   The number of subscribers to pay-TV in Italy

Companies	1996	1997	1998	1999	2000
Tele + [a]	770 173	927 181	1 113 792	1 354 876	1 557 991
Stream	–	–	89 373	385 640	668 568
Total	770 173	927 181	1 203 142	1 740 516	2 226 559

Note: [a] The TELE+ subscribers for 1997, 1998 and 1999 include cable subscribers who numbered 2180 in 1997, 3141 in 1998 and at 30/9/99, 3508.

Source: AGCM, case No. 8386, Stream-Telepiù, and case No. 10716 Group Canal-Stream.

Table 5.3   Estimated value of sports rights (millions of euros) (soccer and other sports)

	2000	2001	2002	2003	Value of rights purchased per subscriber
Germany	1105	1235	1470	1300	35.8
United Kingdom	1035	1145	1345	1180	48.16
Italy	670	720	830	725	33.72
France	535	580	655	570	23.17
Spain	325	350	390	340	25.35

Source: AGCOM Report 2003 p. 73. The value of rights per subscriber was computed by dividing the value of sports rights by the number of households with TV.

of sports rights per subscriber was 307 euros. Stream's losses were 359.9 million euros in 2000 and 484.6 in 2001. Telepiù had an average revenue per subscriber of 348.7 euros and an average cost of sports rights per subscriber equal to 198.4. Telepiù's losses were 429.9 million euros in 2000 and 344.5 million in 2001. As most pay-TV costs are fixed costs, a small number of subscribers does not allow companies to recover costs.

It is very likely that the cause of financial difficulties of pay-TV companies in Italy was the limited size of the market rather than the cost of sports rights. We do not have data on sports rights costs only for Pay TV but we know (Table 5.3) that the value of sports rights for the total TV sector in Italy, on a per subscriber basis, is about average of that of other European countries.

Telepiù's and Stream's financial difficulties pushed the former in 2002 to buy Stream. A reduction of competition and increase in the number of subscribers were seen as a possible remedy to Telepiù's financial problems. This acquisition was allowed by the Italian antitrust authority (AGCM) in 2002. The AGCM placed some obligations on Telepiù to reduce the possibility that the broadcaster could exploit its monopoly position in the pay-TV market. In 2002 negotiations began, between Sky and Canal Plus, for the acquisition of Telepiù. The transaction was settled in April 2003. Sky was obliged by AGCM and by the EU Commission to give other operators access to its platform and to its content. The access price was based on retail minus principle.[7] Sky was also obliged to waive exclusive rights on football matches and films on platforms others than Direct to Home (DTH). The aim of this obligation was to ensure that sporting events, one of the key drivers of digital pay-TV television would not be available exclusively.

The entry of Sky on the pay TV market affected the number of subscribers and the subscription price. Revenue per subscriber increased by 20 per cent in 2003 over 2002, and the number of subscribers increased by 11 per cent over 2001. Therefore Sky revenues increased by 33 per cent in 2003 over 2001. In 2004 the number of subscribers to Sky increased and by the end of November it reached 2.7 million. In the meantime in Italy, as in other European countries, there was a decrease in the price of sports TV rights that benefited Sky's balance sheet.

The entry of Sky led to a monopoly in the pay-TV market but strengthened competition across the whole Italian TV market as a whole. Negotiations for access to content began with the cable operator Fastweb and for access to the platform with the sports producer Gioco Calcio (an association of football teams). These two parties were quite weak[8] and therefore their impact on the structure of the pay-TV industry was negligible.

With the introduction of DTT in 2004, stronger competitors – MEDIASET and TV7 – entered the pay-per-view market. For 110 million euros MEDIASET bought exclusive DTT rights to cover the soccer matches of Juventus, AC Milan and Inter (the teams with the largest following in *Serie A*) for the next three years. For 32 million euros TV7 bought exclusive DTT rights to cover the soccer matches of central and southern Italian teams (Fiorentina, Roma, Bologna, and so on). These prices are much lower than those paid by Sky to football clubs for exclusive DTH rights – 400 million euros for one year. The lower prices allow MEDIASET and TV7 to sell prepaid cards at a very reasonable price (three euros per MEDIASET event and two euros for TV7).[9]

In the short run the impact of the entry of these new competitors in the pay-per-view market is surely positive, since it allows consumers who are not Sky subscribers to see events that they would not otherwise see. Total welfare increases because a larger market is served. In the long run, if the effect of MEDIASET's entry in the pay-per-view market is to decrease the growth rate of subscribers to Sky, the effect on competition between plat-forms will not necessarily be positive. In Italy, competition between plat-forms is much weaker than in other industrialised countries and hence events that weaken such competition are not positive. This point will be analysed in the sections below in the context of the effect of digitalisation on competition in the Italian TV market.

ITALY'S TRANSITION STRATEGY AND FREQUENCY ASSIGNMENT POLICY

Governments have adopted different transition strategies from analogue to digital terrestrial TV that vary considerably in their priorities, instruments and timing. Such policies create different rules of the game for the reorgan-isation of the sector. One of the most important aspects of these policies is the question of licensing. In the United States the US Congress authorised the distribution of additional broadcast spectrum to each TV broadcaster so that the new DTT service could be introduced while simultaneously con-tinuing with current analogue broadcasts. In Britain only the BBC received sole control of a DTT multiplex. The other incumbents were instead forced to share capacity on two other multiplexes. Room was also made for licens-ing new operators. The US strategy made a clear choice to guarantee the survival of free local TV, privileging incumbent local stations in the alloca-tion of DTT licences. The goal has been to extend the organisation of the industry around small local oligopolies in the digital era. In Britain the strategy was designed to change the structure of the industry in favour of new entrants and restrain dominant operators. In Italy the strategy has privileged continuity over reform. The transition strategy adopted is such that the industry's structure in analogue TV will not change in the digital world. Migration from analogue to digital will be led by the two main market incumbents despite the cost in terms of forgone spectrum efficiencies and media access.

Italian governments have often stated that the switch from analogue to digital TV will increase the number of channels at regional and national level. This is true if we consider the number of licences that will be allocated at the end of the transition period, which is scheduled for the year 2006.[10] The Italian Frequencies Plan (statement of AGCOM no. 15/03/CONS) has resolved on 18 multiplexes for the digital TV services, twelve SFNs (single frequency networks) and six MFNs (multi-frequency networks) that allow regional variations. Each multiplex could carry at least three digital chan-nels of the same quality and coverage as the current analogue channels. Thus, at the end of the transition period it will be possible to have at least

54 channels with a coverage of 80 per cent of the territory and 90 per cent of the population. These channels will use much less spectrum than used by analogue channels. At the end of the transition period competition in the TV digital sector will no longer be restrained by spectrum scarcity.

There are two large obstacles that must be overcome to attain true competition for TV digital services. The first is that, as shown, in Table 5.1, the two main incumbents (RAI and MEDIASET) have 80 per cent of the towers and frequencies in the Italian analogue market. These two companies, when the switch to digital is complete, will need between 10 and 20 per cent of such frequencies to provide the same number and quality of programmes with DTT as they broadcast today. Therefore MEDIASET and RAI should surrender frequencies, although past experience shows that Italian governments have been ineffective at obtaining the surrender of frequencies when the law required it (that is Law 249/97). The second obstacle is the strategy adopted by Italy to implement digital TV. The 1998 frequencies plan approved by AGCOM freed up four frequencies to launch four digital terrestrial channels that would allow nationwide coverage. However, a different strategy was chosen with the creation of *network* providers that are in charge of the multiplexes.

Multiplexes were offered to applicants who could obtain the licence to be a network provider, but no broadcast spectrum was allocated to them. Therefore the network provider could use frequencies already owned, if it were an existing terrestrial broadcaster, or it could buy frequencies on the market. Most of the frequencies available on the market were those owned by the hundreds of local TV channels, many of which had financial difficulties. They found it more profitable to sell towers and frequencies than to make investments to implement digital TV on the local market. Existing terrestrial broadcasting also needed to buy frequencies to introduce the new DTT service while simultaneously continuing with their current analogue broadcasts. The lack of a plan in 1975–1990 to assign frequencies to users has resulted in a highly inefficient allocation of the spectrum (with interference problems) that continues today. In some sites existing terrestrial broadcasters have excess capacity and have no need to buy frequencies to offer both the new DTT service and analogue programmes. Act no. 13137 of the Italian antitrust (AGCM) hypothesised that RAI, with 90 plants and frequencies in excess of the needs[11] could offer a potential coverage of digital services to 74 per cent of the population. RAI with another 459 plants in excess could offer effective coverage of 70 per cent of the population. However, the same Act responded favourably to the purchase of extra plant and frequencies by the RAI. A similar hypothesis of excess capacity was made in respect of MEDIASET, but in this case too AGCM authorised the acquisition of new frequencies and towers. The problem is that in some

*Table 5.4   Degree of coverage and market structure of TV platforms in
2004 in Italy*

Platform	Coverage	Market structure
Terrestrial analogue	>99% of the population	Collective dominance by two oligopolists
Satellite analogue	100% of the population (theoretical)	Monopoly
Satellite digital	Large metropolitan areas	Monopoly
Cable	In theory, 80% of telephone lines	Monopoly
ADSL		Dominant oligopoly with a market share higher than 80%
DTT	>50% of the population	Four groups own five multiplexes

places there is no excess capacity and even large broadcasters must buy frequencies to cover them.

To date, five DTT multiplexes have been licensed, all to existing terrestrial broadcasters: the RAI has two multiplexes, MEDIASET has one, Telecom Italia has one, and Holland Co-ordinator, which has financial links with MEDIASET, also obtained one. The government hypothesis is that by 2006 there will be four or five multiplexes that will cover 70 or 80 per cent of households. The first goal of Law 112/04 has already been reached: by the end of 2004 the DTT RAI trials had covered 50 per cent of the population. One of the RAI multiplexes will allow national coverage without regional variation (SFN). Obviously, the Italian government strategy does not encourage the entry of new competitors on the network providers market, as Table 5.4 shows.

ITALY'S REGULATORY RESPONSE TO VERTICAL INTEGRATION IN THE DTT MARKET

Law 66/2001 and the subsequent media laws tried to shape the market structure of the DTT television industry (see Art. 5 of Law 112/2004). The laws identify three kinds of TV operators: the network provider, the content provider and the operator that provides administrative and technical services for digital TV including conditional access and a subscription management system. There is a need of different licences to operate in cable, satellite, DTT platforms. Furthermore in DTT platform there is a need of a licence for each stage of the TV vertical chain. In the two other platforms, cable and satellite is enough for only one licence. The network provider owns the network infrastructure and supplies broadcast transmission capacity. The content provider ensures the provision of programmes.

Granting a licence does not imply an assignment of frequencies, which AGCOM will do by a different administrative act. The previous section explained how frequencies are actually assigned. At least in the transition period, the chances of a newcomer operating as a network provider are very low due to spectrum scarcity. Hence the market for the DTT platform downstream of network providers will be dominated by few firms: the incumbents in the TV analogue market. Two of these companies, RAI and MEDIASET, already dominate this market. Theory suggests that vertical integration may be used to facilitate the strategic practice of exclusionary behaviour. An integrated firm could deny a rival access to necessary inputs for the purpose of gaining or maintaining monopoly power. Vertical integration can raise prices of intermediate and final goods and harm consumer welfare. In programming and distribution in the TV industry, vertical integration could have these effects. Law 112/2004 implicitly recognises the possibility of the practice of market foreclosure and obliges network providers to fulfil the following requirements:

(a)   network providers must provide open access to their platform on fair, reasonable and non-discriminatory terms;
(b)   network providers using a digital terrestrial platform are required to offer 40 per cent of their capacity to companies not owned by the owner of the network provider;
(c)   network providers integrated upstream in the production and programming of contents (where the network provider and content provider are vertically integrated) must establish separate companies, even if they could continue to have the same ownership. This requirement does not hold for cable and satellite TV companies.

Furthermore, Law 112/2004 requires that all companies in the digital communication TV services industry have separate accounting for the different services they produce.

Requirements for companies that are only in the upstream market (the provision and production of content) are weaker than those for companies in the downstream market. Content providers are required to offer access to content at market prices and non-discriminatory terms (art. 5f). This requirement does not apply if the rights are exclusive.

The above requirements increase transparency and could allow the regulator and the antitrust authority to improve their ability to verify the existence of anticompetitive practices (exorbitant final and access prices, discrimination against an upstream company, and so on). On the other hand, Law 112 allows vertical integration between programming and distribution. We could expect vertically integrated firms to follow the strategic practice of market foreclosure. Chipty (2001) examines the vertical relationship between programming and distribution in the US cable TV industry.[12] The paper provides systematic evidence that integration does result in some degree of market foreclosure.

Hence the above obligations do not seem adequate to ensure development, in Italy, of a competitive DTT market. They are weaker than those imposed on incumbents in the telecommunications sector. Only the obligation on vertically integrated firms to establish two different companies for upstream and downstream markets is not applied to dominant telecommunications operators.

In Italy there are two vertically integrated companies RAI and MEDIASET that are dominant in the analogue TV market. In the transition period, extra spectrum is not available for new competitors, and infrastructures used in the analogue market could be used to become a network provider, with a modest amount of investment. The two companies also have enough money to develop new content for the digital market. We could expect these two companies to use their dominant position in the analogue TV service market, and in the network provider digital market, to seek to exclude new content providers.

However, Laws 66 and 112 state that each network provider must assign 40 per cent of its capacity for DTT trial to other broadcasters. This rule, according to the Italian government, will allow an increase in competition in the content providers' market. This rule cannot be substantially enforced when a company has strong commercial links with other companies in the national and international TV market. Two MEDIASET channels are broadcast in the digital trial by the network provider Holland Co-ordinator. The antitrust authority (AGCM) in a recent act (No. 13770, 16 November 2004) observed that a large obstacle to competition in the TV advertising market is the great influence of Fininvest on the other private incumbents Telecom Italia and Holland Co-ordinator.

Our conclusion is that the Italian media laws are inadequate to increase competition in the digital service markets. The goal of achieving a genuinely competitive market can be achieved only with ownership separation between network and content providers at least for digital terrestrial television. Theory suggests that vertical integration may have a number of efficiency-improving effects that ultimately lower prices, improve product quality and increase consumer welfare (vertical integration increases information and reduces transaction costs, eliminates double marginalisation, internalises service and quality externalities),[13] but in the Italian case the negative exclusionary effect will outweigh the efficiency effects.

The recent acquisition by MEDIASET of the exclusive rights on DTT to cover the soccer matches of Juventus, Milan and Inter (the teams with the largest following in *Serie A*) for the next three years will further decrease the opportunity to have a genuine competition in terrestrial TV in the future. Only ownership separation between network provider and content provider can reduce incentives for exclusionary behaviour and allow genuine competition in the content market.

THE REFORM OF MEDIA OWNERSHIP RULES

The new media law 112/2004 changed the media ownership rules of Law 249/77 in a direction much more favourable to the main private incumbent MEDIASET. The key political significance of the change in rules established by law 112 is that regulations have been calibrated to prevent both RAI and MEDIASET being charged with passing the 30 per cent limit of Law 249/97. The Italian antitrust has already started investigations against RAI and MEDIASET for passing the limit.

The old limit prohibiting any company from owning more than 20 per cent of total national networks, with a maximum of two, is replaced with the rule prohibiting a company from owning more than 20 per cent of total programmes broadcast by the digital terrestrial platform.[14] This limit will come into effect when the frequency assignment plan is accomplished. The number of broadcast programmes when the plan is achieved will be higher than 36. Thus the limit of channels that a company will be able to broadcast will be at least seven. Today Rai and MEDIASET each own three channels.

However, until the plan is implemented, there is another ownership limit that replaces the Law 249 threshold of 30 per cent of market share of total revenues of the sector to which the broadcaster (terrestrial, cable, satellite) belongs. The new ownership limit is 20 per cent of total value of SIC (integrated communication system) sector. This sector is the sum of the following subsectors: press, trade directories, radio-television, movies, other advertising, sponsorship, and so on.

The total value of the SIC sector was about 26 billion euros in 2003. Twenty per cent of this value is 5200 million euros; the total TV advertising market was about 4123 million euros, and the total TV market was about 6165 million euros. A TV operator could be a monopolist in the TV advertising monopoly and still not exceed the 20 per cent limit of SIC value . It could also have more than 85 per cent of the total private TV market and not exceed the limit of 20 per cent of total SIC revenues. Such possibilities are made difficult by a more flexible rule, that does not allow the establishment of dominance in any of the submarkets that are part of the SIC.[15]

Law 112/2004 repeals the parts of Law 249/97 that establish antitrust boundaries but states that the establishment of dominance is not allowed in SIC and in any of the submarkets that are part of the SIC. It defers the implementation of the rule to European competition law. Competition law does not prohibit dominance in itself, but prevents abuse of dominance; Law 112 prohibits dominance in itself in SIC markets. The criteria to assess dominance are those set by European Competition law.

The European Commission's definition of dominance (EC, 2002) denies (section 3.1/75) that a share of 25 per cent must result in dominance; a share higher than 50 per cent yields a rebuttable presumption dominance and a share around 40 per cent makes dominance quite likely.[16] In this last case, proof of a dominant position must be supplemented by other structural factors. Law 112/04 specifies such factors: level of barriers to entry, degree of competition in the television market, degree of efficiency of the firm, and other indices concerning the diffusion of TV programmes, of media products, movies, and so on. Table 5.5 shows that RAI and MEDIASET had a market share in the total TV sector, in the last three years, higher than 30 per cent and lower than 40 per cent. Thus, if one considers the total television market (one of the SIC subsectors) market share, in the Commission's decision making practice, gives no evidence of a single dominant position must be proved. Therefore the proof of a dominant position by Mediaset

Table 5.5   Broadcasters' market shares of the total Italian television market (%)

	2001	2002	2003
RAI	42.2	41.8	39.5
RTI(MEDIASET)	34.9	34.0	34.3
Telepiù	11.8	12.8	6.8[a]
Stream/Sky Italia	3.8	4.4	12.2
Network 7 (Telecom)	1.4	1.3	1.6
Others	5.9	5.8	5.5

Note: [a] The Telepiù data refer only to the first half of 2003.

Source: AGCOM, Act No. 326/04/CONS.

and RAI must be supplemented by other structural factors. On the other hand a possible and, I think, correct interpretation of Law 112 is that dominance must be computed with reference to relevant markets as defined by antitrust commissions. Traditionally the national antitrust authority and the EU Commission have segmented the TV sector into different relevant markets. In case No. COMP/JV.37-21/03/2000 – BSkyB/Kirch Pay TV, the Commission states (paragraph 25), 'The fact that subscribers are prepared to pay considerable sums for pay-TV indicates that the latter is a distinguishable product with specific utility'. If we exclude pay TV revenues from total revenues of the TV sector the market share of the RAI is 47.6 per cent and that of MEDIASET 41.3 per cent. On the other hand if we consider only the TV advertising market the share of MEDIASET in 2003 was 64.4 per cent. Furthermore such share has increased over time. Therefore MEDIASET share give evidence of a dominant position. The share of RAI is 28.5 per cent and therefore there is also a clear evidence of collective dominance[17] by RAI and MEDIASET.

Law 112, from one angle, goes back to the concept of dominance in European competition law, but is ambiguous in following the approach of the European Commission in defining the relevant market on which dominance is computed. On the other hand, Law 112 does not take account that the concept of dominance could include collective dominance. A recent lawsuit clarified that the expression 'one or more undertakings' in Article 82 of the EC Treaty means that a dominant position may be held by two or more economic entities which are legally and economically independent of each other (Papadias, 2004). In theory, this interpretation could allow the actual structure of Italian TV sector to change. In practice a finding of collective dominance is quite difficult to prove.[18] The conclusion of this section is that, with Law 112, the Italian government has given up any attempt to regulate media ownership.

THE ROLE OF THE PUBLIC BROADCASTER IN THE DIGITAL ERA

The relaxation of spectrum scarcity does not necessarily ensure that programmes that are beneficial for society as whole will be offered, if they are undervalued by individuals or generate external benefit. A public broadcaster could provide them. The question is whether the public broadcaster should also offer commercial programmes funded with advertising, rather than solely education and information programmes funded by licence fees. In Italy this latter model has been followed with results that were not considered satisfactory by the Antitrust commission.[19] Programme diversity

was too low, with the public broadcaster duplicating commercial programmes rather than serving special interest groups. The quality of programming was low.

The excess of capacity in frequencies and plants has been one of the causes
that has blocked entry and has allowed the formation of a duopoly in Italy.

AGCM also found collusive behaviour between the public broadcaster
RAI and MEDIASET. A different model with the RAI financed only by
licence fees, broadcasting only two channels and freeing up parts of the
broadcast spectrum to increase the number of competitors might have been
a much more efficient choice from a welfare point of view.

Law 112 does not change the organisational model of public broadcasters and introduces a further complication: partial privatisation of the RAI.
There is a contradiction between the claim of the Italian government that
digital TV is a great opportunity for Italy which facilitates convergence of
the existing communications infrastructures (broadcasting, telecommunications and computer technology) and enhances competition across services and platforms, and its defence of the role of the public broadcaster.
Government regulation of terrestrial television was generally promised on
the notion that the natural limitation of the electromagnetic spectrum
required careful government scrutiny of broadcasting in order to ensure
that this public resource was used to the benefit of all. By using the spectrum more efficiently, digitalisation relaxes spectrum constraints on the
number of channels. Thus digital technology has revived questions about
the legitimate role of the state in regulating terrestrial broadcasting.[20]
Many social scientists think a market–based system will not work fully to
supply public interest programmes, and therefore there is ample room for
public TV. Such programmes must continue to be financed by licence fees,
but there is no need for public TV to finance commercial programmes with
advertising resources. Digitalisation reduces, rather than increases, the need
for the public operator to broadcast commercial programmes. This is not
the approach of Law 112, which divides the public operator into two divisions, one that will supply commercial programmes and will be financed by
advertising and subscription, and one that will offer public interest programmes and will be financed by licence fees. The divisions will have separate budgets as Law 112 forbids use of resources from licence fees to fund
commercial programmes. AGCOM will monitor the public broadcaster's
actions. This solution is not very efficient. We have shown elsewhere (Del
Monte, 2004), analysing configurations of viewer preferences, that a DTT
is not necessarily a solution to market failure in the TV market and that a
public broadcaster could be required to satisfy specific programme types.
On the other hand we have shown that it is not efficient for a public broadcaster to produce commercial TV. The only effect of the entry of a public

broadcaster into commercial TV is to duplicate programme types and therefore decrease total welfare.

A further important aspect of the Italian broadcast reform is the partial privatisation of RAI. Partial privatisation will allow a reduction in the public deficit without central government losing control of the RAI. The most probable scenario is the issue on the market of 30 per cent of the shares of the RAI, but Law 112 places a limit on the maximum number of shares that any private shareholder can own at no more than 1 per cent. Agreements between private shareholders on the election of the members of the boards of directors are also forbidden. The total number of seats is nine: three are elected by private shareholders, four by the Italian parliament, and two by central government (through the Treasury). This distribution of seats allows the parliamentary majority easily to retain control of the RAI.

The effects of the new public governance arrangements of the RAI on total welfare could also be studied starting from the results of the literature on mixed markets. One of the results of this literature is that the effect of such a market structure on total welfare will depend very much on the objective function and on the behaviour of the public firm. The literature assumes that public firms maximise social welfare (the sum of consumer surplus and profits made by the firm) while private firms maximise their own profits. De Fraja and Del Bono (1989) show that, in the context of a quantity – setting oligopoly, welfare may be higher when a public firm is a profit maximiser rather than a welfare maximiser. Thus privatisation of a public firm may improve welfare even without improving the efficiency of the public firm. If on the other hand public firms maximise social welfare the best result is reached, in terms of total welfare, when there is Stackelberg competition and the public firm maximise the welfare as a Stackelberg leader. The behaviour of the RAI and MEDIASET on the Italian TV market have been more that of tacit collusion than Cournot competition.[21] Only in this last case is it possible to show (Baraldi, 2003) that partial privatisation increases total welfare. In this case the public company will maximise a weighted average of welfare and profits. The only possibility that the new corporate governance of the RAI will lead to an increase in public welfare, following the mixed markets literature, is that unlike in the past the RAI and MEDIASET will compete and not collude.

Thus, on the basis of the theoretical literature on mixed markets and an analysis of meeting consumer preference, and in light of the past behaviour of the RAI and MEDIASET, the choice of the Italian government on the strategic role and corporate governance of the RAI will not improve welfare.

CONCLUSIONS

Our conclusion is that the future of digital television in Italy will be decided by politics rather than technology. The Italian media laws are not capable of increasing competition in the digital service markets. The goal of a genuinely competitive market can be achieved only with structural regulation. The main focuses of structural regulation policy have to be:

(a) ownership separation between network and content providers, at least for digital terrestrial television;
(b) reassignment of frequencies to network providers in such a way as to increase entry;
(c) complete privatisation of the commercial part of the RAI.

Without such interventions the classic rules of conduct regulation (mandatory access to networks and content, and so on) will be unable to foster competition at least in the DTT market. Rules of conduct may be adequate in a country where there is competition between platforms. Unfortunately the limited diffusion in Italy of platforms competitive with DTT makes such rules inadequate.

In other countries the existence of established cable and satellite networks facilitate the transition to DTT, since it becomes necessary for only a minority of the population to migrate to DTT. In Italy where analogue terrestrial broadcasting is still predominant, the switchover process is very complex, involving the replacement or upgrading of millions of TV receivers and the replanning of thousands of transmitters. Without structural regulation the transition to digital, if the switch date of 2006 is to be respected, will greatly penalise consumers in terms of investment and subscription costs without substantial advantages for new and better services. The alternative, more in the tradition of Italian TV policy, will be to postpone the switch date without substantial change in the structure of the TV industry.

NOTES

1. Pressure from the European Commission is not likely to have a great effect on the structure of the Italian TV industry. Thus Levy (1999) shows that the analogue model of regulation adopted in France, Germany and England has not been very greatly affected by EU level broadcasting regulations.
2. The Final Report of the Working Group on and Digital Territorial Television in EPRA countries shows that in 2003 Italy had 520 local terrestrial channels, Germany 50, the UK 19, Sweden 0, Holland 115 and Finland 3.
3. Another major problem of the Italian television market was that while theoretically 12 licences were granted at national level, some companies could not broadcast, because

they did not have the frequencies required. Europa 7 is one of the companies that had the licence but was unable to obtain frequencies at national level since they were occupied by MEDIASET's Channel 4.

4. Law 223, as all subsequent laws regulating the broadcasting industry in Italy, followed the policy of imposing much tighter limits on advertising on the public than on private broadcasters.

5. Surveys show that in Italy the audience of pay TV is different from that of a free-to-air audience. The high income class represents 56.3 per cent of the total pay-TV audience, and 37.7 per cent of the total TV audience.

6. Phillips Global Media.

7. Case N. Comp/ M. 2876- Newscorp-Telepiù 2/04/2003. The main obligations imposed by the EU Commission were:

- Sky platform must grant all interested third party operators access under fair, transparent, non-discriminatory and cost-oriented conditions.
- Sky will keep separate financial accounts regarding its activities as a provider of technical services for each technical service separately.
- Access to Sky content to all interested third parties, at a fair reasonable and non-discriminating basis, with a price based on retail minus-principle.
- Sky could bid for exclusive film and sports rights only on DTH platforms.
- Sky must waive exclusive rights on films and football matches on platforms other than DTH.
- Sky must offer Premium channels on a non-exclusive, non-discriminatory and unbundled basis, based on retail minus principle to third parties that could distribute such programmes on platforms other than DTH.
- Sky must forgo entry in the DTT market.

8. The cable TV sector is not well developed in Italy. Fastweb has an optical-fibre network in Milan and Rome and it is expanding its infrastructure to other cities, but its coverage is still limited. In June 2004 it had 151 000 TV subscribers. Fastweb is also a Telecom operator and is able to offer TV, broadband services and telephone services.

9. With one million DTT decoders sold, and about 700 000 prepaid cards sold by MEDIASET and TV7, the pay-per-view market is still small.

10. Italy and Finland are the EU countries that have the earliest analogue switching date: 2006. But there is some doubt that Italy will be able to fulfil this timetable.

11. RAI could cover the same share of population and could offer the same quality of analogue services with less spectrum relative to the total it uses now.

12. The Cable industry is divided into a large number of distinct local markets, and cable operators sell to consumers normally in an exclusive franchise area.

13. For a discussion of these effects see Perry (1989).

14. The 20 per cent limit is lowered to 10 per cent for firms belonging to the telecommunication sector that command more than 40 per cent of telecommunication revenues. Only one firm in the telecommunications market has such share, Telecom.

15. Law 112 proscribes until the year 2010 the acquisition or the start-up of newspapers by TV broadcasters that own more than one channel at a national level.

16. The monitoring of the existence of dominance in SIC and in the single submarkets is entrusted by Law 112 to the AGCOM.

17. 'Indagine conoscitiva sul settore televisivo: la raccolta pubblicitaria', p. 50, AGCM, November 2003.

18. In the Italian case such proof is not so hard to seek: the head of the government which controls the public broadcaster is also the owner of MEDIASET.

19. *Indagine conoscitiva sulla raccolta pubblicitaria*, AGCM, November 2004.

20. Armstrong and Weeds (2004) provides an exhaustive analysis of the role of public service broadcasting in the digital world.

21. See op. cit. in Note 18.

REFERENCES

Armstrong M. and H. Weeds (2004) 'Public service broadcasting in the digital world', mimeo, University of Essex.

Baraldi L. (2003) 'Privatization and mixed oligopoly: the case of telecommunications', *Studi Economici*, **80**, 123–149.

Barnett H. J. and E. Greenberg (1971) 'TV-programme diversity–new evidence and old theories', *American Economic Review*, **61**, 89–93.

Cave M. (1997) 'Regulating digital television in a convergent world', *Telecommunication Policy*, **21**(7), 575–596.

Cave M. and C. Cowie (1998) 'Not only conditional access: towards a better regulatory approach to digital TV', *Communications & Strategies*, **30**(2nd quarter), 77–101.

Chipty T. (2001) 'Vertical integration, market foreclosure, and consumer welfare in the cable television industry', *American Economic Review*, **91**(3), 428–454.

De Fraja G. and F. Del Bono (1989) 'Alternative strategies of a public enterprise in oligopoly', *Oxford Economic Papers*, **41**, 302–331

Del Monte, A. (2004) 'Switching from analogue to digital television in a circular city model', mimeo, University of Naples.

Doyle C. (1998) 'Programming in a competitive broadcasting market: entry welfare and regulation', *Information Economic and Policy*, **10**(1), 23–40

EC (2002) 'Commission guidelines on market analysis and the assessment of significant market power under the community regulatory framework for electronic communications network and services', Document 2002/c165/03.

Gabszewicz J., D. Laussel and N. Sonnac (2001) 'TV-broadcasting competition and advertising', Working Paper from Catholique de Louvain-Center for Operations Research and Economics, May.

Galperin H. (2004) *New Television, Old Politics*, Cambridge: Cambridge University Press.

Levy D. A. (1999) *Europe's Digital revolution: Broadcasting regulation, the EU and the Nation State*, London & New York: Routledge.

Noam E. M. (1998) 'Public–interest programming by American commercial television', in E. M. Noam and J. Waltermann (eds) *Public Television in America*, Gütersloh: Bertelsmann Foundation Publishers.

Noam E. M. (2003) 'Corporate and regulatory strategy for the network century', in G. Madden (ed.) *Handbook of Telecommunications Economics*, Volume II, Cheltenham: Edward Elgar.

Owen B. and S. S. Wildman (1992) *Video Economics*, Cambridge, MA: Harvard University Press.

Papadias L. (2004) 'Some thoughts on collective dominance from a lawyer's perspectives', in P. Buigues and P. Rey (eds) *The Economics of Antitrust and Regulation in Telecommunications*, Cheltenham: Edward Elgar.

Perry M. K. (1989) 'Vertical integration: determinants and effects' in Schmalensee R. and R. Willig (eds) *Handbook of Industrial Organization*, vol.1, Amsterdam: North Holland.

Spence M. and B. Owen (1977) 'Television programming, monopolistic competition and welfare', *Quarterly Journal of Economics*, **91**(1), 103–126.

Steiner P. O. (1952) 'Programme patterns and preferences, and the workability of competition in broadcasting', *Quarterly Journal of Economics*, **66**(2), 194–223.

Vaglio A. (1995) 'A model of the audience for TV broadcasting implications for

advertising competition and regulation', *International Review of Economics and Business*, **42**, 33–56.
Working Group on Digital Terrestrial Television in EPRA Countries co-ordinated by AGCOM 2004 – *Final Report*.

6. The development of digital television in the UK

Martin Cave

After a rather shaky start, the penetration of digital television in the United Kingdom had by mid-2005 reached 60 per cent of households, divided among digital cable, digital satellite and digital terrestrial (DTT) platforms. In addition, by the end of 2005 more than 99 per cent of UK households had access to DSL technologies providing broadband using the telephone company's (BT's) copper wires. Although plans to provide IPTV on this platform are still in their infancy, it too will soon be providing additional competitive pressure.

This chapter describes how the UK came to be in this relatively enviable position, and will focus on the close interaction between broadcasting policy and spectrum policy. The first section gives a brief account of the development of TV broadcasting in the UK. The second section outlines the growth and development of digital television. The third section describes the development of spectrum policy including issues associated with digital switchover, for broadcasting, The final section contains conclusions.

THE DEVELOPMENT OF UK ANALOGUE TELEVISION[1]

In common with those in most European countries, Britain's broadcasting system was quickly assimilated into the public sector. The first regular radio broadcasts in 1920 were undertaken by equipment manufacturers which formed a broadcasting consortium in 1922. But, following a government inquiry, the private British Broadcasting Company was in 1927 converted to a public British Broadcasting Corporation (BBC), financed by a licence fee levied on reception equipment, under the direction of the Board of Governors appointed by the Government, but enjoying a high degree of independence from political interference.

The BBC's monopoly of first radio and then television broadcasting was eliminated by the Television Act of 1954, which established regional advertiser-financed stations, which came together to make up the ITV network. This created what a subsequent government inquiry called a 'comfortable duopoly' with the BBC, each operator enjoying a different form of finance and hence having little direct commercial incentive to compete. In 1964, the BBC was awarded a second national TV channel and a second public but advertiser-supported station (Channel 4) with a remit to provide complementary programming which began to broadcast in 1980. A further analogue Channel 5, with a 70 per cent household coverage, started operations in 1997.

Broadcasting transmission for the BBC and for ITV and C4 was undertaken by respectively, the BBC itself and the regulatory body for commercial television, the Independent Broadcasting Authority. These activities were, however, privatised in the early 1990s. Until the creation of Channel 5, the restriction on the number of broadcasters was not the result of spectrum shortage, but a policy decision, driven largely by a fear that a proliferation of competitive commercial broadcasters would degrade public taste.

Under the 1984 Cable and Broadcasting Act, cable networks were licensed in local franchises, and cable would eventually pass two-thirds of UK homes. Forbidden until 1991 to provide telecommunications services, cable was initially fragmented and struggling but after it became entitled to provide telecommunication services, and – later – broadband, customer numbers rose and a process of consolidation into two operators took place throughout the 1990s. But the original substantial investments on cable networks are unlikely ever to be repaid.

Satellite broadcasting began in dramatic fashion in the UK, with a struggle between medium-powered Luxembourg-based transmissions provided by Sky TV and a high-powered service, licensed by the UK regulator, provided by BSB.

In 1990 the companies merged, to form BSkyB following its acquisition of live football rights (described by the company's chairman, Rupert Murdoch, as 'the battering ram of pay television'), and BSkyB built up customer numbers to over 3 million in 1998.

Thus at the dawn of the digital television age, 1998, about 23 million households received analogue terrestrial transmissions, while 3.3 million received pay television from satellite and 2.9 million from cable. The imbalance between satellite and cable is understated by the figures, however, because the vertically integrated BSkyB, as a programme wholesaler, supplied all the cable operators' premium programming. The terms on which it did so were the subject of major disputes in the years from 1995.

THE DEVELOPMENT OF DIGITAL TELEVISION

The decisions by a cable or satellite platform operator to switch from analogue to digital transmission are largely a matter of commercial policy. A cable operator has to make significant network investments but benefits from much greater channel capacity. A satellite operator will have to acquire additional transponder capacity for the inevitable period of duplicated analogue and digital transmission, and may need further additional spectrum for the uplift of programming to the transponders. Customers may have to redirect their reception equipment to a satellite in a different orbital position. But the date and pace of such a change are almost entirely at the company's discretion.

Financial constraints prevented UK cable operators from upgrading their networks to digital until the start of the present decade but the task is now largely completed. BSkyB adopted more aggressive targets, sustained by the offer of free digital set top boxes, to convert the entirety of its customers to digital over a three year period from 1998 to 2001.

The prospect of multi-channel multi-platform television also had a significant impact on the other main major broadcasters, especially the BBC. The Corporation increasingly saw itself as broadcaster which should be present on all platforms. It therefore sought to get access on all platforms not only from its existing analogue channels but also for a range of new digital services. Plans for the latter led to proposals in 1998–1999 that households with access to digital transmission should pay a higher licence fee than analogue subscribers (Davies, 1999). The Government ultimately rejected this in favour of an increase in the licence fee for all viewers to pay for the BBC's digital developments.

The principal commercial broadcaster, ITV, adopted the different policy of not putting its channels on digital satellite – despite a financial incentive to do so.[2] It subsequently changed its policy and became embroiled in a dispute over the charges it had to pay for access to BSkyB's technical platform services. In recent years, such disputes have encouraged broadcasters, including the BBC, to bypass satellite encryption even at the risk that their broadcasts *en clair* might violate the intellectual property rights of their content providers.

If cable and satellite digital services were largely a private matter (but with public consequences) the development of digital terrestrial broadcasting initially required carrying out new spectrum assignments and the issuing of new broadcast licences. This not only created a new multi-channel platform to stand alongside cable and satellite; it also created the possibility of 'switching off' analogue transmission entirely. As a result, the broadcasting market place would change irrevocably, as every household would have multiple channels;[3] second, the spectrum freed by the cessation of analogue transmission (prime spectrum in the most valuable bands, as opposed to less valuable spectrum used for satellite transmission) would become available for other users.

The 1996 Broadcasting Act was the means of creating the new DTT platform (see Goodwin, 2005). It set up six digital multiplexes, each capable of broadcasting four to six channels, initially to about 70 per cent of the population. One was entrusted to the BBC and another to ITV and Channel 4 together. One half of a third was granted to Channel 5. The remaining three-and-a-half multiplexes were available for pay-TV, with a maximum of three multiplexes allowed per firm.

Two bids were received for three commercial multiplexes – one by a consortium of the two largest ITV companies and BSkyB, the second by the largest cable operator. The licence was awarded to the former, but only after the effective exclusion of BSkyB on the ground that it controlled the rival satellite platform.

On-Digital (as it was initially called – later ITV Digital) started operations in late 1998. At its then power level, coverage turned out to be only 60 per cent of the population at most, and its channel capacity fell far short of BSkyB's digital service. It purchased premium channels from BSkyB for resale to its customers, at prices which were subsequently investigated in (and cleared by) a competition enquiry (OFT, 2002). It purchased additional exclusive sports rights at prices which in retrospect seemed excessive. These factors contributed to its bankruptcy in 2002.

The licence was re-advertised, and in July 2002 was awarded to what became known as Freeview. This was a consortium of the BBC, BSkyB and Crown Castle, the privatised transmission arm of the BBC, itself acquired in 2004 by NG Transco, the electricity and gas transmission company.

Freeview was, however, little more than the loose amalgamation of the separate activities of the three organisations: the BBC controls one Freeview multiplex on which it transmits its and others' channels; some channels are supplied by BSkyB; while Crown Castle, the major player, leases space on two multiplexes. Set top boxes are available commercially (at a price of £30–£50), with no subsidy. Almost none of them has a conditional access module permitting pay TV, although there are a few pay TV channels available on Freeview, subscribed to mostly by previous customers of ITV Digital.[4]

Figure 6.1 shows the startling success which Freeview has enjoyed, and how it has catapulted the UK take-up of digital television to being the highest in Europe. It has also confronted BSkyB with a number of challenges. The availability of a multi-channel service with no monthly payment puts pressure on BSkyB's pay TV prices. Additionally a household

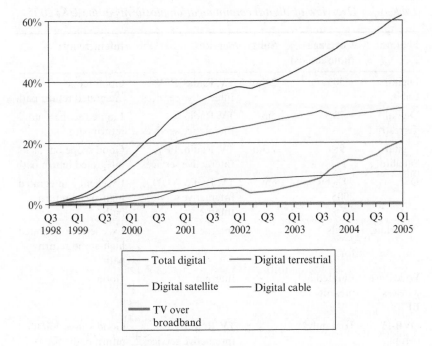

Figure 6.1   The spread of digital TV in the UK

installing a DTT set-top box clearly faces a switching cost in moving to satellite (which would require another box). The success of DTT thus jeopardises BSkyB's prospects of continuing to expand without cutting its prices. For this reason BSkyB has set up a Freesat service which allows a household access at no recurrent fee to a set of channels similar to those on Freeview. Unlike DTT channels, however, these satellite channels are encrypted, and BSkyB make a card available for customers. The impact of this service has so far been negligible and it is not likely to be assisted by the method chosen by Ofcom to ensure universal coverage of public service broadcasting after digital switch-over, which is described below. In September 2005, however, the BBC and ITV announced plans to broadcast an unencrypted free digital satellite service.

This account has concentrated on digital cable, satellite and terrestrial transmission. There are, however other options in play and Table 6.1 summarises them.

Table 6.1   Overview of digital communication platforms in the UK, 2005

Platform	Coverage (households)	Subs	Services	Interactivity
Digital Cable	<50%	2.5m	TV, Radio, PPV, Interactive services	Good scope – integrated return path
Digital Terrestrial	~73%	5m	TV, Radio, Interactive services	Limited scope + no return path
Digital Satellite	~98%	7.3m	TV, Radio, PPV, Interactive services	Good scope but lacks integrated return path
DSL	99% by mid-2005	5m	TV, Radio, VOD, Interactive services, Internet	Excellent – integrated high-speed return path
Powerline	Trials		Internet	Excellent – integrated high-speed return path
Wide area wireless BB	Limited at present		Internet	Good
DVB-H (Mobile TV)	Trial only		TV, Radio, Interactive services	Good – uses 2G/3G return path
3G	Up to 80% (pop.)	~2.m	Voice, Messaging AV stream/ download, Interactive services,	Good

Notes: PPV = Pay per view; VOD = video on demand.

DIGITAL SWITCHOVER[5]

In September 1999 the government first announced its ambition to switch off the analogue TV signal and move to digital transmission. It said that the digital switchover could start as early as 2006 and finish by 2010, although the precise date would depend on the behaviour of broadcasters, manufacturers and consumers.

The government further announced that switchover would not take place unless the following conditions were met:

1.   Everyone who could watch the main public service broadcasting channels in analogue form could receive them in digital.

2. Switching to digital was an affordable option for the vast majority of people.

Currently 98.5 per cent of UK households can receive analogue TV signals for the four main analogue broadcasters. The target indicator of affordability was defined as 95 per cent of households having access to digital equipment before switch-off, generally taken to mean that 95 per cent of households would have adopted digital TV before switchover occurred.

The government launched the Digital TV Action Plan two years after the announcement of the long-run objective of switchover. The plan has successfully carried out a large number of measures in preparation for switchover and has co-ordinated the activities of various parties involved in the switchover, including Government departments, regulators, broadcasters, retailers, manufacturers and consumer groups. The aim of the Action Plan was to make the necessary preparations to allow the Government to make a decision on the exact timing at a later date.

When the switch-off of analogue signals occurs it will be carried out on a region-by-region basis which means that compulsory switchover will take place in different years. ITV regions are likely to be the regions that the public understand most readily. The order in which the regions switch off must take account of technical, logistical and commercial planning. The final order was only recently settled by Ofcom.

The cost of converting receivers will vary according to several factors, including:

- the amount of digital equipment a household already has;
- how much additional equipment the household wishes to continue to use after switchover;
- their platform and equipment choices;
- their service choices; and
- prevailing prices in the year(s) they make their purchases.

However, as for some people switchover will not be voluntary and, for most, affordability will be a major issue, there must be some minimum, one-off cost with which consumers feel comfortable. The cost of conversion for a house with only one TV where a new aerial is not required is estimated at £40–£80, while the cost for a house with two TVs and one VCR may be as much as £80–£160.

Advice on whether any consumers will need publicly funded assistance is currently being sought from the Ofcom consumer panel. In Italy, for example, a subsidy of 40 euros per household is available. The UK government is likely to want to contain the figure by limiting eligibility.

Although half of households have digital TV, the other half are unlikely to get digital as quickly or readily. And unless they can be persuaded to do so, switchover will be more difficult.

In research commissioned by the government, most households without digital TV said they were likely to convert of their own accord within the next few years. However, this leaves some 20 per cent of households who currently intend to remain analogue only. Independent of whether they could receive a DTT signal, three-quarters of this group said they would adopt digital if they knew switchover was imminent, while the remaining quarter (5 per cent of all households) said they would never be willing to convert.

The group least willing to convert to digital TV are not a coherent cluster with clearly defined socio-economic or demographic characteristics. Instead they tend to have a variety of reasons for remaining with analogue TV. A household's propensity to adopt digital television frequently reflects its attitudes towards TV and multichannel TV in particular. Those least willing to adopt digital television tended not to value TV as a medium, or alternatively felt that more TV channels would have a negative impact on society. Some others believed that digital TV had little to offer over and above that offered by analogue TV; while some mentioned issues such as cost and difficulty of use.

Even if consumer take-up of digital were to increase, shutting down analogue TV services will only be a realistic possibility if there is support for the objective of switchover itself. Such support is currently limited. Research conducted for the DTI has indicated that around 50 per cent of people objected to switchover and were suspicious of the government's motives. This suspicion was not confined to those least willing to adopt digital TV – many of those consumers who already have digital TV said they would resent the loss of their analogue services. The same survey found that people accepted that technological advance was inevitable and that at some point digital TV would supplant analogue TV. Nevertheless, a successful implementation of switchover will require an improvement in public support for the goal itself. At present there is very little public recognition or understanding of the underlying case for switchover.

In accordance with normal practice, the Government has presented a justification of the switchover policy in the form of a cost-benefit analysis (DTI, 2005a). This calculates net benefits to the UK in the region of £1.5bn to £2bn. The price of these benefits would be the cost of transition. Around switchover a substantial amount of households would have to convert their primary TV to become compatible with digital signals, and many more households would have to convert secondary TVs and video recorders if they wished to continue to use them. At the same time, the UK would benefit considerably from switchover; which has the potential to transform broadcasting and to offer new services to millions of households.

The analysis evaluates the costs and benefits to the UK of completing a digital switchover involving the switch-off of all analogue signals. This scenario is compared with continuing both the analogue and digital transmissions. The analysis focuses on the quantifiable effects of switchover, including environmental effects. The non-quantifiable effects of switchover such as the public service aspects of the DTV project are not discussed and the distributional aspects of the project are not examined in detail.

The consumer costs of switchover include the net cost of set conversions, which will be necessary for all households not covered by DTV at the time of takeover. It is assumed that set conversions will be done by purchasing a set top box. However, the aggregate cost of purchasing STBs overestimates the economic cost of switchover, as some of these consumers will have been very close to buying into digital even if the switchover was not to take place, that is they value digital TV at some level between the cost of the STB and zero. To model this, it has been assumed that the implicit demand curve for STBs is a straight line from the cost of an STB to zero, and therefore the average valuation by consumers is half the cost of the STB. It is not just primary TVs that will lose functionality at switchover, but all non-digital TVs and VCRs. The average ownership of TVs per household is currently around 2.5, so the costs involved in converting secondary sets are significant and have been included in the analysis. In addition to the cost of set conversions in the form of STBs, the cost of digital satellite conversions for households not covered by DTT after switchover has also been taken into account. Producer costs are the base station infrastructure costs and planning and operations costs.

One of the key consumer benefits associated with switchover is the value of increased DTT coverage to previously un-served areas, areas that it was impossible to reach using digital signals during dual transmission. These benefits are referred to as 'extended coverage benefits'. Consumers will also benefit from the release of 14 channels of clear spectrum when analogue transmission ceases. The economic value of this extra spectrum depends on the use to which it is put: generally it is estimated that it will be of more value if it is used for mobile telecommunications rather than television. However, because of risks and uncertainties associated with the use of using spectrum for mobile telecoms, the analysis is based on the assumption that the released spectrum is used for digital television services.

The key producer benefit from switchover is the saving in costs from decommissioning analogue transmitters, as the cost of running, maintaining and fuelling such sites will no longer have to be borne. It is assumed that any producer surplus arising for the operators of the new services on released spectrum will be competed away.

The cost-benefit analysis shows quantifiable benefits in the region of £1.1 – £2.25 billion in net present value (NPV)[6] terms. Sensitivity analysis gives results that show NPV reducing under some assumptions but remaining substantially positive under most likely combinations of assumptions. The model shows that the outcome in terms of NPV is most sensitive to estimates of the value of extended coverage of DTT services and released spectrum.

A key variable in the analysis is the year of switchover, which can be changed in running the model to show a consequent NPV. Switchover in 2010 shows a positive NPV of £2.25 billion, and the NPV falls by around £250–£300 million for each year that switchover is delayed. The effect of a five-year delay is likely to be around £1.2 billion in NPV terms.

The UK Government has accepted a commitment to ensure a level of convergence of public service broadcast signals equivalent to that currently available with analogue broadcasting. However, this could be achieved by various means – directly by mandating public service broadcasters to transmit in particular ways, or indirectly by placing an enforceable burden on relevant broadcasters to meet a specified availability target, in whatever way they chose. The latter approach would contemplate the possibility of a variety of technologies being employed to provide coverage, DTT, cable, satellite (FreeSat) and DSL or other technologies. Broadcasters with a universal coverage obligation would have an incentive to seek out the cheapest combination from a commercial standpoint; such harnessing of incentives has clear advantages. Moreover, any preference for a single platform inspired by regulator or government would, if accompanied by explicit or implicit state subsidies, raise issues of possible state aid.[7]

Following a lengthy consultation, Ofcom finally decided to mandate DTT as the means of providing universal digital coverage for public service broadcasting multiplexes, although commercial multiplexes were free to make their own choices, so long as coverage did not decrease. Even this prescriptive solution left open a number of trade-offs among the objectives of:

(a)　coverage (raising the level by small amounts) above the current 98.5 per cent available using digital technologies;
(b)　power levels (which determine the number of channels available or a particular multiplex);
(c)　the cost of additional transmitters; and
(d)　the risk that the option adopted would be subject to delays.

The variant which emerged victorious in 2005 was one which allowed more channels to be broadcast by using a particular mode of operation (known as 64/QAM). By using higher power operation at selective sites, it would attain a 98.5 per cent coverage of households at the cost of additional transmitters on the South Coast.

This means that, as digital switchover progressively occurs throughout the UK regions, analogue transmitters will fall silent at each of the current 1154 sites. All of those sites will be used for DTT, in place of the 80 sites currently used, at lower power, to achieve a 70 per cent coverage. The UK would thus effectively replicate its existing analogue networks but with a six-fold increase in capacity.

In September 2005, the switchover period of 2008–2012 was confirmed, together with the outline of a package of measures to support people over 75 or with disabilities, paid for by the BBC out of licence fee income received from all viewers. A not-for-profit company, called Digital UK – formerly known as Switchco – and funded by broadcasters was established to support the transition (DCMS, 2005).

UK SPECTRUM REFORMS AND DIGITAL BROADCASTING

The above account of the development of broadcasting has shown how broadcasting policy has generally driven spectrum allocation rather than *vice versa*. Channels were added as and when broadcasting policy dictated, despite the availability of extra spectrum. Following a period of over twenty years in which broadcasts were simulcast on the UHF and VHF bands, VHF broadcasting ceased in 1985. The emergence of DTT was a highly directed process. The only significant departure was the 'unauthorised' emergence in 1988 of Sky, which used a Luxembourg-based satellite and did not initially require a broadcasting licence from the IBA or a wireless telegraphy licence from the UK Government. But following the merger with its 'approved' rival BSB, BSkyB too came into the regulatory fold.

This subordination of spectrum allocation to broadcasting policy was fully consistent with conventional 'command and control' methods of spectrum management, under which frequencies are allocated to particular services (radar, mobile communication, astronomy, broadcasting) under international agreements, then administratively assigned to particular organisations by national authorities. Because broadcasting is done at high power on frequencies with a wide coverage, the risk of inter-channel and international interference is high. In a densely packed region such as Europe, this necessitated detailed annual planning, which for analogue broadcasting was codified through the Stockholm Agreement of 1961. An equivalent planning exercise for digital will be completed in 2006.

In recent years, however, the spectrum management authorities in several jurisdictions have concluded that 'command and control' mechanisms fail to meet the dynamic needs of innovating sectors where demand grows quickly and unpredictably. In the UK in particular, the Government in 2002 accepted the recommendations of an independent review of spectrum management (Cave, 2002), which proposed greater reliance on market mechanisms to allocate spectrum.[8] The UK Communications Act of 2003 placed on Ofcom, the newly integrated (broadcasting and telecommunications) regulator the duty of seeking optimal use of spectrum, and laid the basis for the introduction of secondary trading and change of use of spectrum, in addition to the auctions of spectrum already used for primary issues. Prior legislation had also permitted the spectrum agency to levy an annual payment for spectrum use by private or public bodies, which became known as an 'administered incentive price'. This was notionally designed to represent the value of the spectrum in an alternative use – its 'opportunity cost' – and to encourage economy and efficiency in spectrum use (Ofcom, 2004b).

Ofcom quickly developed a Spectrum Framework (Ofcom, 2005c) and Implementation Plan (2005d), together with a series of measures to accommodate trading. The strategy envisaged a speedy switch from 'command and control' to market methods, which by 2010 would account for 70 per cent of assigned spectrum (see Table 6.2), another 4–10 per cent being licence exempt.[9]

The levying of charges for broadcast spectrum became highly controversial. As noted above, broadcasting licences assigned via a competitive

Table 6.2   Use of different spectrum management techniques

	Command and control (%)	The market (%)	Licence exempt (%)
(a) Spectrum below 3 GHz			
1995	95.8	0.0	4.2
2000	95.8	0.0	4.2
2005	68.8	27.1	4.2
2010	22.1	73.7	4.2
(b) Spectrum between 3 GHz and 60 GHz			
1995	95.6	0.0	4.4
2000	95.3	0.0	4.7
2005	30.68	61.3	8.2
2010	21.1	69.3	9.6

Source: Ofcom (2005c) p. 36.

tender (even if the competition were not in terms of a monetary payment, but in terms of another quantifiable variable such as speed of roll-out) already involved payment of a spectrum change as part of the bundle. This situation covers ITV and Channel 5. At the re-tender of the digital multiplex licences following the collapse of ITV Digital in 2002, the Government stated there would be no charge for spectrum until 2014.

That left the BBC analogue channels and C4. These broadcasters argued that their obligations to provide universal coverage in practice rules out any flexibility in spectrum use; any charge would thus be a tax, with no efficiency benefit. The broadcasters also stated that they were already incurring the costs of duplicated (analogue and digital) transmission, and would ill afford analogue spectrum charges. Ofcom has so far been silent on how it proposes to deal with pricing for television broadcasting spectrum (Ofcom, 2004b, 2005e: 33).

It follows from Ofcom's strategy that in future, subject to international obligations, there will be no pre-ordained 'broadcasting' spectrum but a variable quantity responding to supply and demand. The obvious test of this policy will be the disposition of spectrum solely to be forced by the digital switchover. Under current international spectrum agreements, its use is restricted to broadcasting, but attempts are being made to change this at the 2006 Regional Radio Conference Planning. As noted above, planning of digital broadcasting spectrum is currently going ahead but in the UK in ways which are designed to maintain flexibility for other users. It is therefore possible that the released spectrum will be auctioned in ways which permit its use for further static digital broadcasting, mobile broadcasting (using a standard known as DVB-H – digital video broadcasting – handheld), mobile communications, or other technologies already in existence or to be developed.

CONCLUSIONS

This review of recent developments in UK broadcasting policy has identified a number of partially conflicting and partially converging trends:

- The key technological change has been the proliferation of platforms now capable of carrying combinations of video programmes, broadband and voice telephony. The result is a major 'widening' of the platform market.
- Consequently, justifications for regulating broadcasting based on market failure arising from limited channel capacity have now fallen

away, as the UK regulator has acknowledged while still finding other arguments for public service broadcasting.

● As a result, the policy focus shifts to the application of competition law and policy, or *ex ante* regulation mimicking the outcome of competition law. In the UK, this has been an arduous and lengthy process, in part because of the complexity of the broadcasting value chain.

● Partially underpinning the process of growing platform competition has been a broader liberalisation of spectrum markets, including use of auctions, secondary trading and change of use.

● However, the need to push through a policy of analogue switch-off, while maintaining seamlessly universal delivery of public service broadcasting channels (that is, all the analogue channels) has forced the government and regulator into a more interventionist stance, over, for example, the universal delivery platform.

This last consideration (combined with continuing support for public service broadcasting and a particular broadcaster – the BBC) has gone against the grain of overall policy. Nevertheless the policy direction in favour of liberalisation is deeply ingrained, and is likely to triumph.

NOTES

1. This material is well summarised by Galperin (2004) Ch. 8; see also Cave and Williamson (1995).
2. The so-called 'digital dividend' meant that advertising revenues earned on digitally transmitted services were not subject to the fixed percentage tax levied on analogue revenues.
3. This would affect, for example, the need for public intervention to support public service broadcasters as a means of overcoming market failure (see Ofcom 2004a).
4. This suits the BBC, which would be more vulnerable to replacement of its compulsory licence fee by a voluntary subscription mechanism if all households had pay-TV capabilities.
5. For a summary, see DTI (2005b).
6. The NPV is the capital sum available today which is equivalent to the expected stream of benefits.
7. This is a particular danger following the Altmark case, in which the European Court specified a need for competitive tendering to be used where possible to finance projects with public subsidies.
8. Market mechanisms had been used previously in the UK to allocate commercial broadcast licences through a competitive tendering process. The object competed for was not a spectrum licence alone but a package involving both favoured access to viewers and the availability of spectrum, conditional upon the performance of specified public service broadcasting obligations (see Cave and Williamson, 1995).
9. Licence-exempt spectrum can be used by anyone abiding by power restrictions. Wi-fi 'hot spots' are a good example of current licence-exempt use.

REFERENCES

Cave, Martin (2002) *Review of Radio Spectrum Management*, London: DTI and HM Treasury.

Cave, Martin (2005) 'Competition and the exercise of market power in broadcasting: a review of recent UK experience', *Info*, **7** (5), 20–28.

Cave, Martin and Williamson, P. (1995) 'The re-regulation of British broadcasting' in M. Bishop *et al.* (eds) *The Regulatory Challenge*, Oxford: Oxford University Press, pp. 160–190.

Davies, Gavyn (1999) *The Future Funding of the BBC*, London: DCMS, HMSO.

DCMS (2005) 'Tessa Jowell confirms digital switchover timetable and support for the most vulnerable' Press Notice 116/05, Department of Media, Culture and Sport.

DTI (2005a) *Cost Benefit Analysis of Digital Switchover*, London: Department of Trade and Industry.

DTI (2005b) *A Guide to Digital Television and Digital Switchover*, London: Department of Trade and Industry.

Galperin, Hernan (2004) *New Television, Old Politics: the Transition to Digital TV in the United States and Britain*, Cambridge, UK: Cambridge University Press.

Goodwin, Peter (2005) 'United Kingdom: never mind the policy, feel the growth', in Allan Brown and Robert J. Picard (eds) *Digital Terrestrial television in Europe*, Mahwah, NJ: Lawrence Earlbaum Associates Publishers, pp. 151–180.

Ofcom (2004a) *Review of Public Service Broadcasting*, Phase 1.

Ofcom (2004b) *Spectrum Policy: Consultation*.

Ofcom (2005a) *Planning Options for Digital Switchover: Consultation*.

Ofcom (2005b) *Planning Options for Digital Switchover: Statement*.

Ofcom (2005c) *Spectrum Framework Review*.

Ofcom (2005d) *Spectrum Framework Review: Implementation Plan – Interim Statement*.

Ofcom (2005e) *Spectrum Pricing: A Statement on Proposals for Setting Wireless Telegraphy Act Licence Fees*.

OFT (2002) *BSkyB Investigation: alleged infringement of the Chapter II prohibition*, No. CA96/70/2002.

7. A perspective on digital terrestrial broadcasting in Japan

Kiyoshi Nakamura and Nobuyuki Tajiri

INTRODUCTION

Digital technologies have greatly changed the environment surrounding the broadcasting industry, resulting in drastic changes to the traditional economics of broadcasting. Digital technological innovation supported by the Morse code-like idea of 0s and 1s and advanced multiplexing technology are blurring the once distinct boundaries between information communication/telecommunications and broadcasting causing these markets to converge. Negroponte (1995) predicted that the 'telephone will become wireless, and the TV become wired'. Innovations in digital technology, however, are progressing at an even greater speed than predicted by this 'Negroponte switch' hypothesis. Now, television can be viewed from the mobile phone, and phone calls can be made through one's cable television connection, increasing the interdependency of fixed (wired) and mobile (wireless) technologies all the more. Moreover the creation of new industries can be expected from such technological convergence.

In Japan there has existed a dual system of commercial broadcasting, which is dependent on advertising income, and public service broadcasting, which is dependent on licence fees. However, new changes to the broadcasting landscape such as digitalisation, the growth of new *subscription broadcast* business models, and the birth of Internet broadcasting as well as mobile broadcasting or 'podcasting' have begun to shake the foundations of this 'cosy duopoly'. From an industrial organisation perspective, it can be said that digital technology is simultaneously giving rise to changes in the broadcasting industry's market structure and market conduct. Horizontal and vertical integration between broadcasters and telecommunications firms including the Internet portal site operators and other strategic behaviour are changing the number and scale of players in the market.

With combined changes to market structure and market conduct impacting on market performance such as profitability, efficiency and fairness, media policy should be developed from an economic perspective in order to design wholesale changes to the broadcasting industry appropriate for the digital age. It must be recognised that the broadcasting market is no longer a 'special market', and is rapidly approaching the status of an 'ordinary market'.

This chapter will first touch on structural changes and the current state of digitalisation in the Japanese broadcasting market, before examining economic policy issues that address the digitalisation of broadcast services.

THE CURRENT STATE OF DIGITALISATION IN BROADCASTING

Digitalisation of Terrestrial Broadcasting

Based on the Japanese government's e-Japan Priority Policy Programme, the digitalisation of terrestrial broadcasting began in December 2003 with the three large metropolitan areas of Tokyo, Osaka and Nagoya. Digitalisation will commence in other areas by 2006, with a final aim of reaching 48 000 000 households. From 24 July 2011, analogue broadcasting will cease, and that analogue spectrum is to be returned to the Japanese government. It is estimated that a total investment of 1.2 trillion yen (about $10 billion) is needed for terrestrial broadcasters to set up digital terrestrial broadcasting facilities throughout the nation.

Tokyo-based commercial networks (so-called 'key stations' such as Nippon Television Network, TBS, Fuji TV, TV Asahi and TV Tokyo) and NHK, sole public broadcaster, had already in August 2005 made more than half of their planned investments in digital facilities. Local stations lag behind due to limited financial resources. The government is temporarily bearing the burden of the costs associated with this digitalisation, as well as providing tax breaks and financial support for necessary investments in digital equipment and infrastructure.

As of August 2005, 8.5 per cent of Japanese have televisions compatible with terrestrial digital broadcasting, according to the Ministry of Internal Affairs and Communications.[1] It is obvious that an early introduction of low-priced products is essential to the fast diffusion of TVs compatible with digital terrestrial broadcasting. The sales of the television sets with digital antennas that utilise liquid crystal display (LCD) and plasma display technologies are expected to rise dramatically, combined with rapidly dropping retail price. The Japanese government predicts that the broadcasting of the Beijing Olympic Games in 2008 and the FIFA World Cups in 2006 and 2010 will spur the take-up of digital television sets by the time analogue broadcasting is terminated in 2011.

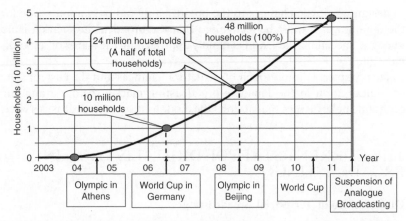

Source: 'Fourth action plan for promotion of digital broadcasting', National Conference for Promotion of Terrestrial Digital Broadcasting, October 2003.

Figure 7.1   The projected diffusion of digital television sets

Issues of Japan's terrestrial Broadcasting Technology

The features of Japan's terrestrial digital broadcasting technology (called ISDB-T) are:

(a)   the signal can be received while mobile;
(b)   the image is high definition; and
(c)   the signal is multi-directional signal, enabling interactivity.

Compared with DVB-T in Europe and ATSC in the US, the identity of each channel is secured, meaning that broadcast signals can be received by mobile devices without interference. In Japan it is now standard for television sets for terrestrial broadcast to come equipped with dial-up/broadband Internet connection functions allowing interactivity.

In the summer of 2005, The Ministry of International Affairs and Communications decided to distribute terrestrial digital television broadcasts simultaneously on Internet-Protocol-based fibre-optic networks. Although it will ensure that digital television services can be viewed even in poor reception areas, it is stirring competition and co-operation between the broadcast and telecommunications industries.

First, since it enables the programmes of Tokyo-based key stations to be directly carried out nationwide, this will endanger the financial resources of local broadcasters that depend heavily on fees from key stations for redistributing key station programmes.

Second, it may make the current regional-specific broadcasting licensing system meaningless, because cross-regional Internet distribution is technologically feasible. Moreover, Internet providers can step in video distribution services. The policy may break down the barrier between broadcasting and telecommunications.

Satellite Broadcasting and Its Digitalisation

Satellite broadcasting can be divided into two categories, 'BS' broadcasting which uses satellites designed for broadcasting purposes, and 'CS' broadcasting, which operates on satellites used for communications. But despite the existence of these categories, technical differences have been almost eliminated between the two.

NHK began fully-fledged BS broadcasting in an analogue format in 1989. WOWOW, the first private satellite broadcasters specialising in movies, music, sports and entertainment, entered the market with a subscription business model in 1991. Digital BS broadcasting commenced in December 2000 and in addition to NHK and WOWOW, several newcomers have provided digital services largely in a high definition format. As of the summer of 2005, more than 10 million households' access to digital BS broadcast services. In the transition to digital, analogue BS broadcasting will be terminated in 2011 as well.

CS broadcasting started in an analogue form in 1992, and was digitalised in 1996, it is a subscription broadcasting platform due to its multiple-channel capabilities. The diffusion of CS broadcasting brought some important changes to the broadcasting market.

First, it provoked a separation between broadcasters who operate the satellites that transmit programmes, and the broadcasters who supply the content to be transmitted. Such a system of separated 'programme-supplying broadcaster' and 'facility-supplying broadcaster' is of great economic significance, as this division between activities that lend themselves to natural monopoly such as the ownership of networks, and competitive activities of production and editing of content, gave content providers more opportunities to get their own channels to consumers, resulting in an expansion of the content market. Later, this separation was also adopted in digital BS broadcasting.

Second, a new market participant was born in the form of a platform operator, which the prevailing broadcasting law does not regulate. The establishment of such an intermediary, which carries out important business functions such as co-ordinating licensed programme-supplying broadcasters, marketing activities, and customer management, is most interesting from the viewpoint of a reduction of transaction costs.

Third, a new regulatory framework was introduced in which CS broadcasting as well as cable television using just telecommunications services is permitted. In this system new entrants to the market are not required to undergo a 'screening of qualifications' by the government as is the case with terrestrial broadcasting, but merely needs to 'register' as a 'broadcaster utilising telecommunications services' in order to provide broadcasting services. Such deregulatory policy can be thought to pave the way for the convergence of broadcasting and telecommunications from henceforth.

CATV and Its Digitalisation

Cable television was first introduced as a countermeasure to the limits of terrestrial broadcasting. Over 17 million households subscribe to autonomous cable television broadcast services in Japan; when adding those who only receive re-transmissions of existing broadcasters' programmes through cables, the total subscription will be more than 23 million. This shows that how far the transition to digital terrestrial broadcasting succeeds will depend on the digitalisation of cable television.

Japanese cable television operators are faced with some issues. First, they are facing the same problem as terrestrial broadcasting of bearing the costs of investing in the digital head-end as the brain of digital broadband technology, which serves as the collection point for the applications required to deliver advanced services such as video-on-demand, interactive television, high-speed internet access and telephony.

Second, further expansion of their networks is the lifeline for cable television operators, hence business tie-ups with other cable operators are becoming a vital strategy.

Third, in the digital age, cable television network plays another role in providing telecommunications services such as Internet access and IP telephony. For cable operators, the connection to other communications networks holds the key to the provision of ISP and IP phone services for 'one-stop shopping', which is becoming an important revenue source.

POLICY ISSUES RELATING TO THE DIGITALISATION OF BROADCASTING

Oligopolistic Structure of the Commercial TV Market and Its Performance

The Japanese commercial terrestrial TV broadcasting market is marked by the existence of five 'key stations'. This grouping into networks is a product of news agreements and the nation-wide transmission of commercials, and furthermore it imposes restrictions on local broadcasting stations' programme production capabilities. The formation of these affiliated groups was necessary to take advantage of network economies. It is said that the income local broadcasters receive from key stations selling fixed timeslots ('network time') as a package and individually through advertising agencies exceeds 25 per cent of their total revenue. The programmes aired during this network time are produced by the key stations. Due to their ability to reach the national market through these networks, the key stations are able to improve their return on investment in content production and in doing so, reduce independent productions by local stations. The result of this is the further increase in the local stations' dependency on the key stations, the reduction of local stations' capability to create independent programmes, and the strengthening of the oligopolistic hold which the key stations have on the market.

In such an oligopolistic commercial broadcasting market it becomes possible for a broadcaster to anticipate its opponent's actions to a certain extent. Assuming that there is competition over programme quality, this can lead to the homogenisation and standardisation of programmes. As implied by the traditional Hotelling's model (Hotelling, 1929), when the principle of minimum differentiation begins to work, programmes become increasingly similar. In reality, the degree of programme homogeneity in the commercial broadcasting market is high; programmes that become hits are often imitated immediately. If the multi-channel environment brought about by digital technologies were to promote a change from broadcasting that reaches all viewers to narrowcasting that targets designated viewers, the economic welfare of viewers, would be improved

Multi-Channel Broadcasting and Public Programmes

While the expansion of multi-channel digital broadcasting increases viewers' freedom of choice supposedly improving their economic welfare, a multi-channel environment does not necessarily guarantee the supply of high-quality programmes. Take the example depicted below in Figure 7.2, based on the Noam Model (Noam, 1998), where revenue and cost are shown in the Y axis and the X axis represent programme quality, ranging in order from low-quality to high-quality. Assuming that advertising revenue and licence fees or subscription reflect the number of viewers, the revenue curve (R) can be expected to show a near-normal distribution centring on the programmes of average quality that all viewers would be likely to watch. Further, production costs (C) are assumed to be fixed. Although multi-channel broadcasting allows the distribution of programmes to specific viewers, it is difficult for commercial broadcasters to provide

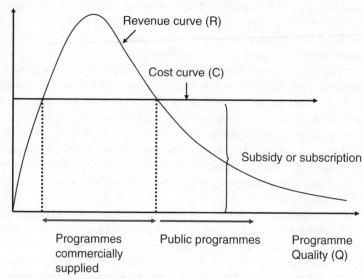

Source: Adapted from Noam (1998), pp. 150, 154.

Figure 7.2   Content quality and public programmes

programmes with low ratings, as they are dependent on advertising and subscription fees. While Q in Figure 7.2 shows programmes whose quality is above the standard bench mark, in order to promote the delivery of such programmes, in other words to overcome this 'market failure', it is necessary to introduce some sort of complementary policies. Herein lies the economic significance of public service broadcasting.

The value of public service broadcasts, like any other public goods or services such as parks and museums, depends on their being used. But the value of choosing not to consume a service now but having the ability to take advantage of it sometime in the future is often referred to as 'option value', which is initially introduced by Krutilla (1967) and later elaborated as a contingent valuation method (CVM) by many economists and public administrators. If many people have the probability of watching a public service broadcast now, there is a chance that the option value is not being considered and that this could result in underestimating the benefits of public service broadcasting.

The kind of organisation through which public service broadcasting is provided greatly depends on a country's culture and traditions. In Japan and European nations, where broadcasting is partly used for cultural and educational purposes, it is often provided by an independent public body. And in Europe, many countries employ a system whereby licence fees are supplemented with advertising revenue. In contrast, in the US, which holds 'freedom of speech' in high regard, the provision of services are left to market mechanisms as much as possible.

As Foster (1992) suggests, revenue sources for public service broadcasting such as tax, subscription or licence fees, donations and contributions, advertising revenue and foundations are often employed, although combinations of these are also possible. Regardless of which method is chosen, no solution exists to address all necessary angles such as the fairness of income distribution, the user-pay principle, consumer sovereignty, and the efficient allocation of resources. Against this backdrop, and in order to promote the production and distribution of 'public' programmes, Peacock (1986, 1996) has proposed for public funds competition that can be used to finance public and commercial broadcasters alike.

While broadcasters are currently under an obligation to maintain a certain amount of educational and cultural programmes regardless of funding methods, as can be observed in Japan's 'programme harmonisation principle' and Britain's public service broadcast obligation, measures to increase 'public' programmes are not only being taken by direct regulations, but also through market mechanisms such as competitive bidding for public broadcasting funds.

Content Provision and Sutton Model of 'Vertical Product Differentiation'

It can be supposed that with the advancement of convergence in communications and broadcasting, the demand for audio-visual content will increase. Furthermore, in examining the degree of fair competition in broadcast markets, it is necessary to consider the economic characteristics of vertical product differentiation of media content.

Generally, product differentiation can be separated into horizontal and vertical. For example, just as anyone would choose an automobile with speed rather than one without, in a situation where all consumers were to make the same evaluation, products would be differentiated vertically. In contrast, the choice between a sedan and a sports car would be made according to consumers' values and needs. News, live sports, motion pictures and other content is in high demand from viewers. Such content is called 'killer content', but in order to acquire the exclusive rights to broadcast this content, an enormous upfront investment is necessary, and once this investment has been made, the risk of being unable to recover this sunk cost is extremely high. It is necessary to examine what effect this will have on market structure.

Traditional industrial organisation theory considers an increase in market size to promote market entry and thus reduce market concentration. Sutton (1991) however, indicated that such an inverse relationship between market size and concentration may not always be applicable in industries where there are large expenditures on advertising and research and development to improve quality. According to Sutton, advertising costs and R&D costs are endogenous sunk costs intended to promote the vertical differentiation of goods and services, and increase the price consumers are willing to pay. As market scale grows, such sunk costs may burgeon, allowing incumbents to exploit their competitive advantage. This indicates that significant sunk costs acts as more of a brake to market entry, not reducing concentration but rather resulting in an inevitably oligopolistic market structure.

Sutton's approach can apply to analyse the market structure for vertically differentiated media content where consumers highly value a higher-quality such as killer-content. Based on Carlton and Perloff (2005),[2] the market structure for vertically differentiated media content is summarised below.

As Sutton suggests,[3] suppose that a consumer surplus of a viewer from a media content is determined by quality and its price. If the marginal cost of producing a high-quality content and prices can be assumed to increase 'sufficiently slowly' as quality rises, the market is likely to be a natural oligopoly, because everyone values the same high quality content so that the provision of new content clearly aimed at niches not served by existing broadcasters becomes difficult. Differing from Hotelling's horizontal product differentiation model (Hotelling, 1929), there is no room to select a certain quality, making new entry difficult. In a word, Sutton's model suggests that there will be no reduction of concentration in a vertically differentiated market, leading to the formation of a natural oligopoly. As Vickers (1985)[4] analysis indicates, whether a market will be prone to natural oligopoly or not depends on viewers' preferences over content and the technological standards that determine the quality of content.

Sutton's model indicates that media content markets may have a natural oligopolistic structure due to their endogenous sunk costs. If fewer subscription-broadcasters hold exclusive broadcast rights to 'killer-content' that the public hope to watch, a problem of fairness may arise between viewers who are able to pay and those who are unable to pay to watch.

Asset Specificity of Content and 'Hold-up' Problem

The production of media content such as documentaries and television series requires upfront costs just as motion pictures do. Content for television however, is often produced especially for a particular broadcaster, meaning the content can usually only be used by that broadcaster. In this way, the investment made for a particular trading partner is called a 'relationship-specific investment'.[5] However, such asset-specificity comes together with the risk that, if for some reason the situation changes, the terms of business may be altered to the advantage of the party with strong bargaining power.

For example, despite a key station commissioning the production of a programme, if for some reason or other the broadcast is cancelled or postponed, payment for costs incurred in production may be delayed or may even not be paid. If an independent production company relies on broadcasters to cover the majority of production expenses, or does not employ renowned actors, they may have no choice but to accept these changes. As the programme has been produced for a particular broadcaster, a problem arises from its asset specificity, meaning it cannot be sold in the open market. While such a state of affairs can possibly be avoided through a written contract, it is impossible to include all unforeseen situations. Due to such incomplete contracts, the potential always exists for the responsibility/liability to fall to one side.

In this way, losses from asset specificity, incomplete contracts, and differential bargaining power due to sunk investments can be explained through the concept of quasi-rent. Here, let's consider the relationship between asset specificity and quasi-rent for general goods based on Besanko *et al.* (2000).

Imagine that, a client decides not to buy a relationship-specific product at the contracted price (P^*). If investment (I) has already been made and is thus sunk, and the producer will continue production so far as variable cost (C) can be recovered, the quasi-rent is thus the difference between selling to the contracted party at the expected price of P^* and the necessity of selling to another party at a lower price (Pm). Here if production volume is I, then quasi-rent (R) can be expressed as follows:

$$R = (P^* - C - I) - (Pm - C - I) = P^* - Pm \qquad (7.1)$$

In other words, this formula shows quasi-rent to be the difference between the expected ex ante profit before a contract is fulfilled, and the expected ex post profit after the non-fulfilment of a contract due to asset specificity. If the quasi-rent were significant due to the high degree of asset specificity, the party with the stronger bargaining power may use the threat of cancellation or suspension of transactions to transfer part of this quasi-rent to themselves.[6]

Such a situation is referred to as the 'hold-up' problem. The hold-up problem is often referred to in relation to so-called 'rights windows' between key stations and independent production companies in Japan.

Further, there is the problem of how long term contractual relations between independent producers and broadcasters ought to be, and who should bear risk and costs. In addition to promoting the efficient distribution of media content, the systematisation of contracts and monitoring of contracts' execution from a view of fair trade are imperative.

Copyright Issues

With the rapid penetration of high speed Internet access in Japanese households, the demand for audio-visual content is continually increasing. At present, however, nearly all audio-visual content is in the form of terrestrial broadcast television programmes. The main reasons for this are that: (a) the processing of rights for secondary use is traditionally not undertaken, and (b) according to contracts, broadcasters retain copyrights, pointing to the fact that its use by the original producers would be restricted. The main cause for the former is that as different copyrights and other related rights for television programmes are vast in number, *ex post* processing costs are extremely high.

The sharing of production expenses and risk in content for television broadcasting in Japan can be divided into three categories:

1. station-produced programmes, created in-house by commercial broadcasters;
2. jointly-produced programmes, where stations collaborate with production companies; and
3. outsourced programmes, where production is contracted out to a production company.

Production companies have such a major role in producing for television to the extent that approximately 70 per cent of programmes broadcast during prime time (or so-called 'golden time' in Japan) period of 7–10 p.m. are said partially or fully to originate from them. Yet despite this close, co-operative relationship between production companies and broadcasters, a more competitive relationship exists between them when it comes to management of the programmes' copyrights and secondary utilisation.

In particular, as it is said that terrestrial broadcasters are in a dominant position *vis-à-vis* the production companies from whom they commission productions, the possibility exists for this market dominance to be abused, creating a problem for fair trade. This involves broadcasters retaining the copyrights for programmes produced for them, or similarly holding the copyright and monopolising the secondary use of programmes. This can be understood as an abuse of market power. Further, for a commissioned production company to have the programme it has produced refused by the broadcaster, or to have its payment reduced would clearly be a violation of anti-monopoly legislation.

In assessing a 'dominant position' in the market, the Fair Trade Commission should take every factor into consideration such as the difference in power between concerned parties, the dependency of the production company and its ability to supply other customers. In Japan, the number of terrestrial broadcasters has been limited by a regulation in the name of spectrum scarcity, meaning that entry into the terrestrial broadcasting market has been restricted. Thus it can be said that a buyers' market situation may easily occur. Ninety per cent of contracted production companies have a capital base of less than 100 million yen, and such a large gap in corporate power between them and the broadcasters leave them apprehensive that contractual terms will be set to the advantage of the broadcasters. So as not to create such a problem, it is necessary to clarify terms, for example using profit sharing. Similarly, such systematisation of contracts is imperative because it promotes the distribution of media content.

However, due to the rapid convergence of telecommunications and broadcasting, it becomes obvious that a handful of licensed broadcasters are difficult to dominate in the business of broadcasting video content, even though they still have a strong voice in the management of copyright. Recently major groups of copyright owners such as actors, scriptwriters and musicians and commercial broadcasters reached an agreement on royalties for programmes distributed on the major portalsite operators such as Softbank, the parent company of Japan's largest portalsite operator, Yahoo Japan.

On the one hand, this sort of tie-up will cause major broadcasters to restructure their business models drastically. Since the Internet providers are not required to have licences, they have the potential to be a strong contender in the oligopolistic market where a handful of licensed broadcasters have been dominated. On the other hand, to obtain smoothly the consent of copyright owners, the security of copyrights should be secured. It is extremely important to set up a reliable system to protect and enforce copyrights.

Relaxation of Media Concentration Laws

As broadcasting plays an important role in shaping public opinion, its political neutrality, diversity and plurality of information and news are highly regarded. Broadcast has been entrusted by society with a unique economic activity, on the assumption that they will refrain from transmitting untruthful, biased information or propaganda. Furthermore, just as in markets for general goods and services, the regulation of market power and freedom of entry are necessary in order to promote fair and free competition. Anti-media concentration laws were established to pursue these political and economic objectives.

Due to technological innovation, however, through convergence of communications and broadcasting markets, the public are continually gaining access to new information distribution channels such as the Internet. This raises the question of just how applicable the traditional argument to restrict media concentration in order to avoid restricting the freedom of speech and opinion. Movements to relax these media concentration laws may provide an opportunity for the local broadcasters through merger and acquisition deals to survive in the digital age.

On the other hand, corporate strategies that try vertically and horizontally to integrate content on its path from production through to distribution have been strengthening. The encirclement of content and resultant raising barriers to entry may possibly lead to an oligopolistic or monopolistic market. Even if the market is not concentrated, it is also possible for consumers' choice to be limited due to lock-in caused by Conditional Access Systems (CAS), Electronic Programming Guides (EPG), and other such technological developments. As market conduct changes in response to digitalisation, it is necessary to examine economically what impact a relaxation of media concentration laws would have on market structure and competitive relations in the broadcasting market.

Japanese media ownership rules were reviewed in an effort to promote the digitalisation of terrestrial broadcasting. The regulations prohibit a firm from holding stakes of more than 10 per cent in each broadcaster in a region or stakes of 20 per cent or more in broadcasters in different regions. Considering the heavy burden of digitalisation investment on the local broadcasters and the possible bankruptcies, exceptional measures have been introduced to allow the approval for mergers in the case of a region adjacent to another region up to two regions during times of financial difficulty.

From the viewpoint of economics, it is problematic to measure the effectives of market concentration. In the West, market dominance is decided by relative market share, but in measuring this, the number of viewers, viewing time, and revenue from advertising and subscription fees should be considered. There is, however, no clear argument in Japan to whether dominance for broadcast media markets should be considered in relation to different genres such as news programmes, and whether it should be assessed or addressed regionally. Furthermore, whether such economic power should be tackled through anti-trust laws or commercial trade laws has yet to be studied.

Spectrum Allocation and Spectrum Charges

Needless to say, radio spectrum is so important a public asset that there is a great need to allocate it efficiently and fairly in order to enhance public welfare. Due to a rapid increase in demand for mobile phones, the shortage of assignable radio spectrum has become a serious issue in Japan.

According to the Ministry of Internal Affairs and Communications' *White Paper 2004*, Japan's information and communications industry as a whole reached 116 trillion yen (about 1 trillion dollars) in the fiscal year 2002. The average growth rate of the sector was 7.1 per cent between 1995 and 2002. Industry-wise the sales of telecommunications services totalled 18.66 trillion yen and those for broadcasting services 3.74 trillion yen in fiscal year 2002.

As for spectrum use, the government anticipates that there will be a great demand for spectrum by potential users such as medicine, publishing, distribution, education, welfare and game related business areas, in addition to security and transportation. Accordingly, it is imperative to implement a drastic policy for reallocating frequencies to meet new demands in wireless access systems and mobile communication systems.

The Japanese government launched in 2002 a study group on policies concerning radio spectrum allocation to examine innovative measures for effective spectrum use. The study group reviewed the current spectrum use fee system from the viewpoint of its ability to encourage efficient use. Currently in Japan, the spectrum fee is charged to cover the common costs of the monitoring service and technical test service for all spectrum users. As a result of a sharp increase in mobile phone users, an unfair burden of cost among the spectrum users has become obvious.

In fact, since the number of mobile phones in Japan has already exceeded 85 million as of the end of 2004, cellular phone companies became major payers of the spectrum use fee. In fiscal year 2004, total revenue from the use of radio spectrum under the current fee system is estimated to reach about 500 million dollars (57 billion yen), of which mobile phone companies will pay 80 per cent, while television stations pay less than 7 per cent.

The study group proposed calculating spectrum fees taking into consideration the economic value of wireless access and mobile communications. Wireless LANs and consumer electronics which are not required to get spectrum licences are exempted from the user charge in order to promote development of those growing industries.

In relation to auction as utilised in the United States and Europe, the study group reached the conclusion that although the auction system is regarded as a way to use efficiently radio spectrum and an important source of government revenues, it could cause financial problems for the licensees and push them to the brink of bankruptcy due to high costs of bidding as experienced in Europe. Moreover, there could be a possibility of cutting off the service for the unprofitable areas and creating vested interests, given that the licences are for 20 years; this would deter the advance of IT industry.

The Office of Communications (Ofcom) in the United Kingdom has been proposing spectrum trading and liberalisation of spectrum use based on Martin Cave's review on radio spectrum management (Cave, 2001). The most important implication which we can draw from UK policy is that Ofcom regards most types of licensees' rights as tradable assets. Obviously there are cases when an administrative approach to the management of spectrum may be appropriate, as Ofcom points out. When we contrast the dramatic advance in digital technology with the serious shortage of radio spectrum in Japan, it is necessary to discuss further various possible mechanisms for effective use of spectrum. This should include spectrum trading.

DIGITAL INNOVATION AND FUTURE BROADCASTING POLICY

Because of limited spectrum and broadcasting's social influence, broadcasting has traditionally been regulated. This regulatory system in which new entrants have been restricted has consequently helped incumbents to sustain their revenue. In industrial organisation terms, this may be seen not as oligopolistic conduct creating dominance, but rather as this market dominance being created by regulation.

However, digital innovation has been promoting multi-channel, high-definition, and interactive functions, and the technological convergence has been advancing the convergence of broadcasting and telecommunications markets. Throughout such dramatic changes in the environment, traditional media policy clearly has its limits, because the assumptions the policy has been based on are more or less weakened and become closer to other markets where consumer sovereignty is fully embraced.

As Vickers (2002) indicates, technological innovation is transforming the broadcasting industry from a 'special' economic sector into a 'general' one. On the other hand, broadcasting is essentially a network industry, and an important strategy for such a network industry is to build further on existing systems. This is different from traditional concepts of competition that deal with 'carving up the pie', rather it aims to first 'grow the existing pie'. On this argument, media policy should switch its focus from market-structure oriented policy to market-conduct oriented policy that creates rules for fair competition.

In relation to competition policy, we should learn from lessons of the past. In the early 1980s, the US airline industry underwent dramatic deregulatory changes. While it is said that the result of this was that 230 companies entered the market only to have 200 of them exit, it was out of such a competitive market that many new corporate strategies were born. Computerised reservation systems, frequent flyers or mileage programmes that offered free flights and locked customers in, and the so-called 'hub and spoke' pattern flight routes were formulated to reduce costs. The structural changes that result from this competition may be nothing more than a move from an oligopoly to a renewed oligopoly, but even in this oligopoly, the potential for competition still exists.

Motokawa (1992), a Japanese biologist, suggests in his book that 'large animals become smaller, and small animals become bigger' on isolated islands if there is no natural enemy. This is referred to as 'the Law of Islands'. And if this sort of levelling out will come into play, animals cannot respond to dramatic environmental changes and will face extinction. Even in the Japanese broadcasting industry, this danger of averaging out exists due to regulation which benefits the incumbents, and in order to avoid this, it is necessary to create a system where the innovative trial of new ideas is possible.

NOTES

1. As of August 2005, the accumulated number of digital terrestrial receivers reached 5.8 million units.
2. Carlton and Perloff (2005), pp. 268–274.
3. Sutton (1991), pp. 70–71.
4. The paper was reprinted on pp. 3–21 in Tim Jenkinson (1996) ed. *Reading in Microeconomics*, Oxford University Press.
5. See in details Besanko, *et al.* (2000), p. 153.
6. See in details Besanko, *et al.* (2000), pp. 153–160.

REFERENCES

Besanko, D., D. Dranove and M. Shanely (2000) *Economics of Strategy*: (2nd edn) New York: John Wiley and Sons, Inc.

Carlton, D. W. and J. M. Perloff (2005) *Modern Industrial Organization*, (4th edn) New York: Pearson and Addison-Wesley.

Cave, M. (1989) 'An introduction to television economics', in G. Hughes and D. Vines (eds) *Deregulation and the Future of Commercial Television*, Aberdeen: Aberdeen University Press.

Cave, M. (2001) *Radio Spectrum Management Review: A Consultation Paper*, Radiocommunications Agency.

Foster R. (1992) *Public Broadcasters: Accountability and Efficiency*, The David Hume Institute: Hume Paper 18, Edinburgh: Edinburgh University Press.

Hotelling, H. (1929) 'Stability in competition,' *Economic Journal*, **39**, 41–57.

Jenkinson, T. (ed.) (1996) *Reading in Microeconomics*, Oxford: Oxford University Press.

Krutilla, J. V. (1967) 'Conservation Reconsidered,' *American Economic Review*, **57**, 777–786.

Milgrom, P. and J. Roberts (1992) *Economics, Organization and Management*, Englewood Cliffs, NJ: Prentice Hall Inc.

Motokawa, T. (1992) *Zo no Jikan and Nezuni no Jikan, Chuko-shinsho* (Time for Elephant and Time for Mouse), Japan.

Nakamura, K. (1999) Japan's TV broadcasting in a digital environment, *Telecommunications Policy*, **23** (3–4), 307–316.

Nakamura, K. (2001) 'Japan's Broadcasting and Telecommunications: Digital Convergence, Market Structure, and Competition,' in Yan-Ching Chao, Gee San, Changfa Lo and Jiming Ho (eds), *International and Comparative Competition Law and Policies*, New York: Kluwer Law International.

Negroponte, N. P. (1995) *Being Digital*, Cambridge, MA: MIT Press.

Noam, E. (1998) 'Public-interest Programming by American commercial television,' in E. M. Noam and J. Waltermann (eds) *Public Television in America*, Gutersloh, Germany: Bertelsmann Foundation Publishers.

Peacock, A. (1986) *Report of The Committee on Financing The BBC*, Chairman Professor Alan T. Peacock DSC FBA, Home Office, UK.

Peacock, A. (1996): 'The political economy of broadcasting,' in *Essays in Regulation*, No.7, Regulatory Policy Institute, Oxford University.

Sutton, J. (1991) *Sunk Costs and Market Structure*, Cambridge, MA: The MIT Press.

Vickers, J. (1985) 'Strategic competition among the few – some recent developments in the economics of industry,' *Oxford Review of Economic Policy*, **1** (3), 39–62.

Vickers, J. (2002) 'Competition policy and broadcasting', in a speech at the IEA conference on the The Future of Broadcasting, UK.

PART II

Content rights and digital broadcasting

8. Legal and economic issues of digital terrestrial television (DTTV) from an industrial perspective

Koichiro Hayashi[1]

INTRODUCTION

This chapter discusses the legal and economic questions associated with digital terrestrial television (DTTV) from an industrial perspective by examining, as an example, the Japanese DTTV service launched at the end of 2003. The author hopes that this analysis provides a common perspective not only to the United Kingdom and the United States but also to other countries that intend to introduce digital terrestrial television services.

LAUNCH OF DIGITAL TERRESTRIAL TELEVISION BROADCASTING[2]

In Japan, the DTV service was started in three distinct major metropolitan areas, namely Tokyo, Nagoya and Osaka, on 1 December 2003, nearly five years later than in the United States and the United Kingdom. The broadcasts were made by Nippon Hoso Kyokai (Japan Broadcasting Corporation, better known as the NHK) and 16 commercial broadcasters. Satellite television services from what is referred to as the 'broadcasting satellites' (BS) and 'communication satellites' (CS) as well as television services from some cable television operators had already changed to digital broadcasts. However, given that almost all households across the country enjoy terrestrial television, the changeover from analogue terrestrial television to digital broadcast is considered to have a much greater impact.

In fact, this historical service made a rather quiet start for a number of reasons. First, DTTV carries almost the same content as the analogue service. Second, the analogue service will continue alongside DTTV until 2011, hereinafter referred to as 'simulcast'.[3] Third, the digital service in the Tokyo area, where more than 30 per cent of all Japanese households are concentrated, was initially only available to 120 000 households located around the Tokyo Tower,[4] with the exception of NHK's general channel.

It is said that the DTTV generally has the following five features.[5]

1. high-quality pictures and sound;
2. data broadcast services that are offered simultaneously with ordinary images and interactive functions;
3. multi-channel programming of up to three channels in a single bandwidth;
4. broadcasting for mobile phones; and
5. server-type broadcast services.

At the current stage, the high-quality pictures and sound are emphasised while other features seem to have yet to become fully available. Amaya (2004) explains that data broadcasting mainly covers independent content that is comparatively easy to produce, such as news and weather information. Actual cases of data broadcasting linked to ordinary programmes include the overall ranking of the Hakone Ekiden Race[6] and the introduction to the course, produced by Nippon Television Network Corporation. Cases in which the interactive functions have been provided include the participation of an approximate 70 000 home digital referees in NHK's year-end song contest,[7] provided via the BS service and the terrestrial service.

The multi-channel programming leads to dispersion of viewers and is therefore inconvenient to commercial broadcasters that provide free-to-air services based on advertising revenue. Even so, some private broadcasters in the Nagoya and Osaka areas have initiated attempts in this area. Late at night, one broadcaster airs two different news bulletins in the same time slot. Another broadcasts two different TV shopping programmes on two channels. Among the future attempts for multi-channel broadcasting, there is a plan to provide a combination of a school education programme, a programme for children and a language learning programme on NHK's educational channel.

Services for mobile phones and the server-type broadcast services will both be launched in 2005 or thereafter. The broadcast services for cell phones once faced a problem with the licence fee for the video encoding technology but this issue has been resolved. With the aim of establishing open operating regulations and standard specifications for server-type receivers, a project to formulate the regulations on the operation of server-type broadcast services was set up in September 2003. Active discussions are now underway as part of the project.

THE SUBSTITUTION HYPOTHESIS

In Japan and in the rest of the world, broadcasters and policymakers tend to think of DTTV as nothing but an extension of the traditional concept of broadcasting. In other words, DTTV is nothing more or less than television broadcasting based on digital technology instead of analogue technology. This idea is hereinafter called the substitution hypothesis.

The following facts are thought to be indicative of the substitution hypothesis.

1. It is believed that the feature of the terrestrial television service in which it has long been regarded as a universal service,[8] whereby it is available to anyone living anywhere without incurring any special cost, should definitely be maintained.
2. It is believed that the terrestrial television service should be distributed using radio waves.[9]
3. It is believed that the analogue television service should be discontinued after a certain period (in 2011 at the latest) and everything should be digitised.
4. To ensure high-quality broadcasting as one of the advantages of DTTV (mentioned on p. 140 (item 1)), each broadcaster is assigned a bandwidth of 6 MHz.
5. No new entrant to the DTTV market alone will be admitted until 2011. For the time being, licences are granted only to existing analogue broadcasters.
6. If images are transmitted using a client-server configuration, services that are close to broadcasting could be provided via the Internet (Noam *et al.*, 2004). However, the challenge has been taken of establishing a framework of server-type broadcasting services.
7. As a long-established tradition in the broadcasting industry, the arrangement where NHK coexists with commercial broadcasters is maintained.
8. For commercial broadcasting, a business model based on advertising revenue is taken for granted.

If digitisation is implemented on the principles described above, it will inevitably resemble a state-run project or a project in a planned economy. The Ministry of Internal Affairs and Communications (MIC),[10] the ministry responsible for broadcasting, prepared the 'Action Plan for the Promotion of Digital Broadcasting' as a basic plan for the collective efforts of the parties concerned to promote the efficient propagation of DTTV. In May 2003, the 'National Conference for the Promotion of Terrestrial

Digital Broadcasting' was set up, participated in by top leaders from a wide range of fields, encompassing broadcasters, consumer electronic retailers, consumer groups, local governments, mass media and economic associations as well as related governmental bodies including the MIC. In this way, the public–private joint promotion system has been established.

As a successor to the Action Plan, the National Conference formulated the Fourth Action Plan for the Promotion of Terrestrial Broadcasting in October 2003. It set the digital broadcasting receiver penetration targets at 12 million units owned by 10 million households, including digital television sets, set-top boxes for DTTV, recorders and PCs. The deadline for achieving this target is the 2006 FIFA World Cup Germany.

Concerning cable television, only a little over 10 per cent of subscriber households now enjoy digitised services. But the number of such households is rising more quickly than the total number of households enjoying DTTV. All cable television services are expected to have been digitised by 2010, which is one year earlier than the discontinuation of analogue television.[11] And as an extension of this growth, it is considered possible to propagate DTTV to all 48 million households in Japan, although this full penetration is merely proposed as a goal.

To reach the goal, it is necessary to have DTTV devices sold at the following average rate (Amaya, 2004):

- approximately 4 million units per year in the three-year period from 2004 to 2006;
- approximately 12 million units per year in a three-year period from 2007 to 2008; and
- approximately 21 million units per year in a three-year period from 2009 to 2011.

No matter how enthusiastically the government promotes digitisation, however, Japan is a market economy. Achievement of the national goal is by no means guaranteed. The NHK reports that the subscriptions to its BS digital television services eventually surpassed 5 million at the end of December 2003. Since the beginning of the services, broadcasters have been calling for the marketing of low-priced receivers. The circumstances surrounding DTTV are identical.

Let us now make a comparison in actual prices between three different 32-inch televisions, namely CRT, plasma and LCD models, from the same manufacturer. The price ratio is as follows.

$$CRT : Plasma : LCD = 2 : 4 : 5.5$$

The CRT model is much less expensive than the other two. A little more than 200 000 yen would be sufficient to purchase a CRT television whereas the LCD model would cost more than 500 000 yen. A 25-inch CRT model is priced at nearly 100 000 yen. The current marketing strategy oriented to the wealthy members of society may result in slow penetration of digital television instead of rapid propagation (Amaya, 2004).

While some are sceptical about the future of DTTV, others are more optimistic. Figure 8.1 indicates the trend in the penetration of monochrome television and colour television. They both followed a typical S-shaped curve that we find in textbooks. In addition, it took about ten years for the radio, monochrome television and colour television to reach a penetration of 80 to 90 per cent. The optimists stress that the plan is quite feasible if DTTV follows these precedents.

THE PARADIGM SHIFT HYPOTHESIS

For the purposes of this chapter, the view that is diametrically opposed to the substitution hypothesis is referred to as the paradigm shift hypothesis. According to this view, DTTV is different from conventional broadcasting although it is still one kind of broadcasting.

The argument can be summarised into the four points as follows (summarised from Hayashi, 1999; 2000).

1. The conventional broadcasting service is premised on a vertically integrated model that combines conduit with content. However, in the digital era, they should be treated separately.[12]
2. Specifically, it is not desirable to continue to use the airwaves as a means of sending content. A range of transmission channels, including wired and wireless channels, terrestrial or satellite, should be used.
3. To respond to the diversity in conduits and receivers, transmission using Internet Protocol (IP), which is the de facto standard on the Internet, should be implemented.
4. If any households are still unable to enjoy the broadcasting programme after introducing an unrestricted combination of conduits as described above, some public measures should be taken to cope with the problem, although it is considered unnecessary in reality.

Broadcasters and policymakers around the world deliberately or unconsciously neglect this hypothesis. However, in view of the present situation in which telecommunications networks are being fully absorbed into the Internet or the IP networks, it is natural to think that broadcasting networks,

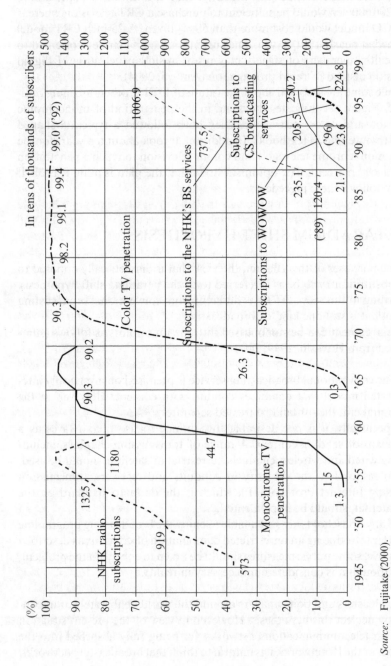

Source: Fujitake (2000).

Figure 8.1 Trends in the penetration of radio and television services in Japan

or at least the portion that corresponds to the conduit, will eventually be incorporated into the IP networks.

In fact, in the field of the telecommunications networks, the substitution hypothesis was dominant for a while after digital networking was proposed. However, the concept of an Information Network System (INS) suggested by Yasusada Kitahara, who was also an advocate of digitisation, is now seen as having served as a precursor of the paradigm shift hypothesis against his intention (Kitahara, 1984).[13]

However, to formulate the idea into a business plan, it is requisite, though not sufficient, that the following expression prove true:

$$Ra \geq Rd \qquad (8.1)$$

Where, Ra is the cost of updating the existing equipment based on the analogue technology with new analogue technology and Rd is the cost of updating it with digital technology.

Here, the existence of critical mass (CM) cannot be overlooked. It is a challenge that is commonly faced by network related businesses. CM is a hurdle that is perceived at an early stage of network development. Any network that has cleared this hurdle can develop autonomously while any network that fails to clear the hurdle is destined to collapse. The existence of CM was discovered as early as the early 1960s by Rogers (1962). Hayashi (1992) demonstrates that it applies to telecommunications networks. Noam (1992) develops this empirical revelation into a theory, and suggests that the development is as portrayed in Figure 8.2.[14]

The diagram indicates that between 0 and n_1, which is the CM point, network development requires some assistance given that $P=AC > U(n)$. Thus expression (8.1) has to be split into two expressions: (8.1$'$) and (8.1$''$).

Before reaching CM,

$$Ra \geq Rd - S \qquad (8.1')$$

where S refers to (external) support.

After CM, there is no difference from Expression (8.1)

$$Ra \geq Rd \qquad (8.1'')$$

Above, S was defined as (external) support to imply that it may be either pure external support or internal reserves. In fact, when Nippon Telegraph and Telephone Public Corporation (commonly known as Denden Kosha) was privatised to become Nippon Telegraph and Telephone Corporation (NTT) soon after Kitahara's suggestion on INS, it was fortunate to enjoy

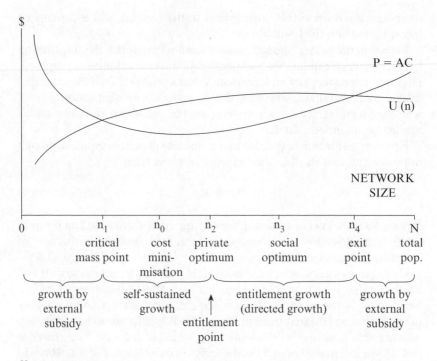

Figure 8.2   Network Tipping

Notes:
AC: Average cost.
$U(n)$: Utility of the whole network when the number of subscribers to the network is n.

Source: Noam 1992.

high profit earnings and a healthy cash flow. Under the conditions where both Expression (8.1′) and Expression (8.1″) are true since S is zero, NTT was able to proceed with the digitisation of its networks on its own.

Promoted in accordance with the substitution hypothesis, digitisation unavoidably came under review in the 1990s. The digital transmission of computer signals on communication circuits had been implemented since the early 1960s, but in the 1990s, there arose a new demand for the Internet. Computing and communications were becoming inseparable. This phenomenon generally referred to as 'convergence', raised a new real managerial question as to whether to choose telephone switchboards or routers.

At that time, telephone carriers were about to introduce digital switchboards, D-60 and D-70 according to the NTT specifications, to replace the analogue ones. They complied with the circuit switching systems but did not support the packet switching that is suited to computer-based communication. Therefore, they were sequentially introducing asymmetric transfer mode (ATM) switches to support packet communications.

In the world of the Internet, service providers were even reluctant to guarantee end-to-end connectivity. They thought that it would be sufficient to provide bucket brigade style transmission on a best effort basis. (The term, best effort, may sound positive but, in fact, the connectivity is not guaranteed.) Then a router was developed for relays. It had such limited functions that it cost considerably less than a telephone switchboard.

At that point, telecom carriers were urged to make a choice: whether to continue to follow the line according to the substitution hypothesis or to convert to the paradigm shift hypothesis. Specifically, this meant whether to introduce routers instead of ATM switches to integrate into IP by discontinuing the cell method. However, long-established carriers including NTT refused to make the conversion. It goes without saying what is now happening because of this.

In the course of evolving into a global de facto standard, the Internet Protocol eventually took over the world of voice communication. In addition, rapidly growing mobile phones are reducing the demand for conventional fixed phone communications. Traditional carriers around the world are now in crisis. It appears that they have no choice but to shift to IP telephones or risk losing their customers.

This implies that although dependence on the substitution hypothesis as a business model may guarantee a solid start, it may sometimes be overtaken by the paradigm shift hypothesis. This explains why Noam (1992) used the term, network tipping in Figure 8.2. He implied the existence of exit point (n_4), where the network will collapse due to its success. We should understand that the lesson learned from the history of telecommunications will definitely apply to the world of broadcasting.

QUESTIONS ASSOCIATED WITH THE SUBSTITUTION HYPOTHESIS

If the paradigm shift hypothesis is adopted, there are almost no given conditions to be considered and it is possible to study a range of scenarios, rather like starting with a blank sheet. Although it is unrestrictive, it tests the system design capabilities (Hayashi and Ikeda, 2002). On the other hand, reliance on the substitution hypothesis automatically sets limitations. There is less difficulty in elaborating on a system and the degree of freedom is lower. The broadcasting industry is bound by a range of unique restrictions, and the constraints imposed by these conditions stand out.

To the best of my understanding, questions concerning DTTV and its dependence on the substitution hypothesis in the industrial context include the following:

1. Questions over the transition to DTTV
    (a) Will broadcasters be able to afford the investment in digitisation?
    (b) Will broadcasters be able to find any new source of income from their investment in digitisation?
    (c) Who should bear the cost of analogue-to-analogue conversion for the simulcast during the transitional period?
    (d) Does DTTV have to be made available across the country?
    (e) At what household penetration rate may the analogue service be closed down?
    (f) Is it necessary to make any rescue efforts if a broadcaster goes bankrupt?
2. Questions peculiar to broadcasting and already known before DTTV
    (a) Does the service area of broadcasting have to be based on prefectures?
    (b) Is the broadcasting service a universal service? (Is the terrestrial service a universal service?)
    (c) In broadcasting, must the hardware and software be combined?
    (d) To administer the radio waves, is it necessary to target the frequency and to be in the form of examination (beauty contest)?
    (e) Is it inappropriate to use radio waves for any purpose other than those specified in the licence application?
    (f) Is it necessary to implement the principle of the decentralisation of mass media?
    (g) What is the attitude towards Japan's unique system of BS services?
3. Questions that are emerging in connection with the antitrust policy
    (a) What is the attitude towards the must-carry rule?
    (b) Abuse of the 'superior position' in programme production contract agreements.
    (c) Position of broadcasters in copyright legislation.

These questions are brought up because the conventional concept of broadcasting based on the substitution hypothesis is considerably remote from the common perception in the market. A fixed idea about broadcasting acts as a kind of myth to dominate those in the broadcasting industry. Table 8.1 breaks it down to compare it with the author's own view.

As there is not enough space to provide an item-by-item explanation, this article gives only a brief summary.

Table 8.1 *The industry's myth and the author's view on the concept of broadcasting*

	Category	Myth	The author's view
Radio waves	Direct wireless reception	Programmes are transmitted using radio waves and viewers directly receive them	Reception is possible using wired transmission or satellite transmission as well
	Scarcity of resources	Radio waves are among scarce resources	Radio waves have some scarcity, but gradually diminishing
	Frequency administration	It is the best method for radio wave administration	There are means of space controlling radio waves other than administering the frequency
	Assignment	Beauty contest	Auction, licence and so on
	Radio wave usage fee	The fee is restricted to the cost of administration	An amount that is commensurate with the market value should be charged
System	Dedicated receivers	Services are enjoyed using dedicated televisions	Services can also be enjoyed using personal computers, while shared displays and mobile terminals will emerge in the market
	Unidirectional communication	Viewers only receive programmes	Interactive communication will be introduced
	Analogue	All services are based on analogue technology	Digitisation is expected to be completed in 2010
	Viewing places	At home	At home, at the office and even in transit
Public nature	Public nature	It clashes with the principle of market forces	The public nature is found in the market
	Universal service	All broadcasting must be available all over the country	Terrestrial television is the only service that has to be a universal service
	Diversity of views	Every prefecture must have more than one station	Given that Japan has a small land area, it is sufficient to ensure diversity at a regional (multi-prefectural) level

Table 8.1 (continued)

Category		Myth	The author's view
	Freedom of broadcasting	The freedom is protected by the government	Freedom projected by the government is not genuine freedom
Management	Hardware and software	It is the more favourable to put hardware and software under integrated management	Whether to integrate or separate the management should be determined on a case-by-case basis
	Restriction on reruns	In principle, all programmes should be broadcast one time only	Given that programmes are content, the channel on which they are aired and the number of times they are broadcast should be freely decided
	NHK and commercial broadcasters	The current structure of coexistence is preferable	The NHK should be privatised so that all broadcasters can compete on an equal footing
	Key stations and local stations	The current network structure is preferable	Local stations should select independence, merger or transition to subcontractors
	Participation in the management of BS stations	Broadcasters have a mission to ensure different channels	Decisions should be based on feasibility under independent management
Copyright		It is natural that broadcasters have a certain right over the programmes aired	It is unacceptable that someone has the neighbouring right when they do not contribute to any creativity
Segregation between communications and broadcasting		Segregation is possible (the current arrangement is preferable)	It is impossible to draw a line of demarcation. Establishment of a new law should be studied
Regulatory body		Uncertain (some may be emotionally in favour of establishing an independent commission)	The competent part of the MIC that corresponds to the former MPT should be reorganised into an independent commission.

1. We will start with the administration of radio waves. Many people think that the peculiarity of broadcasting lies in the use of radio waves that are assigned nearly free of charge to facilitate business as in the past. However, the era in which only the frequency is administered is ending. Technology is now available in which output regulations alone inhibit interference if radio waves are used freely. The era in which there is a revolution in radio waves may come at some time, although few people in the broadcasting industry, let alone among the public in Japan, still believe it to be true. (Ikeda, 2003).[15]

2. Next, we will consider the stance adopted towards the system of broadcasting. Those in the industry tend to be overly persistent about the current structure. They are particularly unwilling to see TV receivers integrated with personal computers. But the times are steadily changing. Reception on mobile terminals will be widespread. When one believes that TV should be enjoyed at home, one is imprisoned with a fixed idea. In the future, the demand from offices should be taken into consideration.[16]

3. Let us move on to the idea of public nature. Nothing is vaguer than this concept. Denden Kosha pursued both 'services to the public and business profitability' (as prescribed in Article 1 of the former Nippon Telegraph and Telephone Public Corporation Law) and ended without attaining either. In view of the author's experience in the promotion of privatisation, the only solution would be to attain a public nature through business (or fairness through efficiency) (Hayashi, 1998). In consideration of this, the conception of the universal service should be substantially narrowed.

4. We will next discuss the form of management. Some argue that the distinct coexistence of NHK and commercial broadcasters represents the ideal structure, but it is nothing but complacency in the convoy system. From the users' perspective, it will be more beneficial to privatise NHK to put it under the same conditions as commercial broadcasters so that they can compete in the market.[17] The layer-by-layer conception in favour of separation between hardware and software is nowadays common in industry. The broadcasting industry should make an autonomous decision on whether to adopt it.[18] Those in favour of integration are probably captured by the myth that radio waves are quality assets, as mentioned in (1) above.

5. Finally, with respect to the integration between communications and broadcasting, those who believe that it is possible to cope with the future using the status quo may simply be reluctant to lose the so-called convoy system.

In short, the broadcasting industry is still far from a normal industry, because of its special position although, in principle, it consists of private companies, with the exception of the NHK (which is a special corporation prescribed by a special law). There is the misunderstanding that the digitisation of terrestrial television can be implemented with the convoy system maintained.

Based on the fundamental perceptions discussed above, of the major questions listed above, the evaluation in the subsequent part of the chapter will focus on the investment in digitisation (p. 148, 1 (a), (b), (c) and so on), the principle of decentralisation of the mass media (2 (f)) and the 'must-carry' rule (3 (a)).

INVESTMENT IN DIGITISATION AND MANAGERIAL COSTS

From the viewpoint of the substitution hypothesis, the investment in digitisation is the mere replacement of existing equipment. The two conditions below are implicit prerequisites for the investment in digitisation.

1. Replacement with digital equipment is less costly than that with analogue equipment, as in Expression (8.1').
2. Investment in digitisation does not result in any particular growth in revenue.

If people in managerial positions have replaced the hardware at their own will, they must have considered these two preconditions. However, digitisation was actually triggered by a comment made by a top official of MPT in spring 1997 (Hirai, 2003a, 2003b). It was by no means a project voluntarily proposed by business managers.

It is understandable, then, that a company under ordinary passive management, would feel like asking for subsidies for the reason indicated by Expression (8.1'). In addition, the circumstances surrounding terrestrial television services in Japan involve some peculiarities. They are:

1. As discussed above, terrestrial television has been positioned as one of the so-called universal services.
2. Furthermore, Japan has more than 10 000 relay stations. This is because TV services have been supposed to be carried using terrestrial transmission despite the mountainous terrain.
3. On the assumption that the simulcast is provided, it costs nearly one trillion yen to build new relay stations for DTTV on the same scale as those for the analogue services.

4. On the other hand, forced universality of terrestrial television produces an adverse impact. As a result, many cable television operators
   in Japan are small.[19] Digitisation of these systems is time-consuming
   and expensive.
5. In addition to the CS services, there are the anomalous BS services.
   Commercial broadcasters cannot escape from the myth that they must
   operate BS digital services as well (see Table 8.1 above).
6. The relationship between key stations and provincial terrestrial stations
   is superficially confined to distribution of news bulletins and programmes carrying nationwide advertisements. In fact, however, provincial stations are under the control of their respective key stations in the
   distribution of content as well as in human resources and capital.
   Provincial stations produce an average of only 13 per cent of the programmes they air (Hayashi, 2002).[20]
7. In the course of deliberation on digitisation, it was found that DTTV
   creates interference with the current analogue UHF services and that
   it is necessary to take temporary measures to eliminate the problem. It
   is called analogue-to-analogue conversion and it specifically includes
   replacement of antennas and alterations to the channel settings.

It is said that the total cost of digitisation exceeds one trillion yen. It has
turned out that the analogue-to-analogue conversion alone costs 180
billion yen to solve the problem mentioned in 7. Given that it would not
have the effect of increasing advertising revenues, the National Association
of Commercial Broadcasters in Japan (NAB) initially estimated that all
local broadcasters except those operating in the Tokyo, Nagoya and Osaka
areas would operate at a loss and would not recover until 2010 (NAB,
1998). According to the revised forecast that reflected a subsequent slow-
down in the Japanese economy, all broadcasters, including those responsible for the three metropolitan areas, would operate at a loss and their
deficit would increase after a temporary upward trend (NAB, 1999 and
Research Institute of NAB 2000). In an 'ordinary' industry, the digitisation
project under these circumstances would be perceived as suicidal and result
in a shareholder-initiated lawsuit.

In the broadcasting industry, however, management insists that digitisation is a national policy. This means that they request financial support
from the government without realizing their managerial responsibility. It
has eventually been decided that the cost of the analogue-to-analogue conversion will be covered by the increase in radio spectrum user fees. The
major sources of the radio spectrum user fee revenues include the mobile
terminal test charges. It looks as if mobile phone users provide financial
support for TV viewers, or mobile phone carriers provide financial support

for broadcasters. Possibly in response to a prick of their conscience, MIC also increased the radio spectrum user fees paid by TV broadcasters, but the structure of the subsidisation remains unchanged.

PRINCIPLE OF THE DECENTRALISATION OF MASS MEDIA[21]

In the Broadcasting Law, Article 2-2 provides for a basic broadcasting plan to facilitate 'enjoyment of freedom of expression through broadcasts by as many nations as possible by creating opportunities for broadcasting for as many of them as possible'. The principle of decentralisation of the mass media has been established based on this policy.[22]

Any person who wishes to obtain a radio station licence for broadcasting purposes (Paragraph 2, Article 6 of the Radio Law) is required to comply to the essential standards for the establishment of broadcasting radio stations (an MIC ordinance)[23] as stipulated in Item 4, Paragraph 2, Article 7 of the Radio Law. Paragraph 1, Article 9 of the MIC ordinance limits the number of broadcasting stations controlled by a single person according to the principle in which one person may own and control no more than one broadcasting station or broadcast programming operation. However, this principle has four exceptions as prescribed in Paragraph 2: (1) operation of a television station and an AM broadcasting station serving the same area, (2) relay stations set up by private broadcasters in their respective service areas, (3) community broadcasting that requires special treatment and (4) the case in which special treatment is required for the propagation of broadcasting.

Here, 'control' is defined as (1) ownership of at least one fifth of the voting rights of an ordinary terrestrial station or more than one tenth if the controlled station covers the same service area, (2) dual appointment of more than one fifth of the officers and (3) dual appointment of any officer or full-time officer with representation rights, according to Paragraph 6 of the same article. Furthermore, Paragraph 3 of the article bans any single operator from running or controlling three businesses in radio, television and newspapers.

Out of fear that the digitisation of terrestrial television may put provincial broadcasters in a situation of financial adversity, MIC partly amended the principle of decentralisation in 2004, as summarised in Table 8.2. It is noteworthy that the amendment dealt with business failures, an area that had been previously regarded as taboo.

However, this minor reform is not sufficient to solve the major problem with transition to DTTV. If broadcasting was perceived as an industry,

Table 8.2 *Details of the relaxed principle of decentralisation of mass media*

Issue	Institutional revision
1 Integration among local stations covering different service areas	A. If their service areas are adjacent, the restriction on equity position applicable to integration among a maximum of seven areas is relaxed from 'prohibition of ownership of one-fifth or more of the voting rights' to 'prohibition of ownership of one-third or more of the voting rights'.
	The two nearest service areas separated by the sea are regarded as adjacent.
	The relaxation is not applicable to any station serving the Tokyo Metropolitan area.
	B. If either (1) or (2) below is applicable to the adjacent areas, the restriction on equity position is not applicable.
	1. One of the areas subject to integration is adjacent to all the other individual areas.
	2. The group of areas is specially defined as that which has close regional unity (for example six prefectures in the Tohoku region, seven prefectures in the Kyushu region plus Okinawa Prefecture).
	The two nearest service areas separated by the sea are regarded as adjacent as in the preceding case.
	The exemption is not applicable to any station serving the Tokyo, Osaka or Nagoya Metropolitan areas.
2 Exceptional treatment in the case of financial problems	In the event of any of the following conditions, the restriction on the equity position and the restriction on officers of a station that concurrently serve as officers for another station is not applicable.
	1. A decision has been made to start corporate reorganisation proceedings pursuant to the Corporate Reorganisation Law.
	2. A decision has been made to start civil rehabilitation proceedings pursuant to the Civil Rehabilitation Law.
	3. Asset deficiency continues in the last two consecutive years with ordinary losses in the last three

Table 8.2 (continued)

Issue	Institutional revision
	consecutive years including the fiscal years of asset deficiency.
	Exceptional treatment includes:
	(a) Permission of investment from a key station*[1] to a local station*[2].
	(b) No more than full ownership authorised (Merger unapproved).
	(c) The regular principle of decentralisation of the mass media is reapplied to the equity position after the recovery of the financial position.

Notes:
*[1] Key station: Refers to a station that serves the Tokyo Metropolitan area.
*[2] Local station: Refers in principle to a station that serves a particular prefecture

Source: Ministry of Internal Affairs and Communications (2004).

ex post regulation in accordance with the anti-trust legislation would suffice. It will be essential to change the perception that the mass media must be regulated ex ante.

In this respect, Funada (2002) deserves credit for his approach of reviewing the decentralisation policy based on the philosophy of the Anti-trust Law. He has, however, been so deeply involved in the regulation of broadcasting industry that his incremental method that suggests unhurried deregulation on the primary basis of the conventional 'broadcasting order' instead of the 'market' has limitations. We should realise that it is time to make the Broadcasting Law consistent with the Anti-trust Law in accordance with an overarching concept of competition policy.

THE 'MUST-CARRY' RULE

The must-carry rule is a regulation set up by the US Federal Communications Commission (FCC) that obliges cable television broadcasters to retransmit terrestrial television services targeting their respective service areas.[24] It means that every cable television broadcaster must carry all television services in the television market for the area in which it is located. If it fulfils this requirement and still has any empty channels, it may redistribute other broadcasting services within certain limits. (This is called the 'may-carry' rule.) If programmes on network affiliated stations and independent stations are included, the range of options is much wider.

In Japan, the Cable Television Broadcast Law has neither a 'must-carry' provision nor a 'may-carry' provision. Retransmission has been handled by mutual agreement between broadcasters. There was no special problem in the age of analogue, but in the era of digital broadcasting, there are more empty channels. Cable television services will be more appealing if they carry programmes on terrestrial TV stations outside their service areas.

On the other hand, terrestrial broadcasters in the market may suffer a loss of viewers. There arises a serious clash of interests. At present, this issue is particularly conspicuous in the Osaka Metropolitan area, where residential zones adjoin each other and the prefectural borders are intricate. In this region, due to the characteristics of radio waves, cross-border redistribution is implemented and is today established as a vested interest.

In fact, the trade price of the programming depends on negotiations, irrespective of whether it complies with the 'must carry' or 'may-carry' rule. The essence of the question should not be whether or not the programming is carried, but how much it is carried for.

In addition, if it has traditionally been possible to retransmit outside the service area and if people have enjoyed watching any programming that is retransmitted in this manner, terrestrial television licensing on a prefecture-by-prefecture basis is questioned. When satellite television services, which refer to the BS and CS services, came into being, the approach based on prefectural boundaries was already obsolete, as pointed out on p. 148 (2 (a)) above. Today, when even server-type broadcast services are technically feasible, the location of the origin of transmission is irrelevant from an industrial perspective.

PURSUIT OF A CULTURE MODEL – A SUBSTITUTE FOR A CONCLUSION

These facts have gradually indicated that it is necessary to understand the need for a fresh model for television culture after the programme production processes change from analogue to digital, instead of regarding DTTV merely as the replacement of equipment.

The philosophy espoused by a major advocate of this view, Yutaka Shigenobu,[25] supposedly reflects the following aspects (Shigenobu (2004) and others).

1. Television has a mission to improve the quality of programmes rather than to manage the conduit. (It seems that he regards the separation of hardware and software as prerequisite.)
2. No one knows what culture will be created from the production of digital content.
3. Whether or not a new culture can be produced depends on the abilities of the creators. Creators must give up their conventional dependency on broadcasting stations to become independent.
4. There has been a tendency in copyright arrangements for the neighbouring rights of broadcasters to be so overestimated that they hold all the rights as the suppliers of funds. The creators themselves should be proud of being the authors, as discussed on p. 148 (3 (b) and (c)).

Shigenobu focuses solely on content production. He is either uninterested in the industrial aspect or is deliberately neglecting it, although he understands the relevance of the industrial angle. However, he goes beyond the substitution hypothesis to support the paradigm shift hypothesis. He and the author see the issue from quite different standpoints but we both make almost the same point. And that seems to be indicative of something.

NOTES

1. I finished this chapter by the end of January 2005. Therefore, some important policy changes, such as the permission by MIC to transmit DTV signals by using the optical fibres and IP technology (Information and Communications Council's recommendation in early August 2005), have not been reflected in this chapter.
2. In Japan, there used to be two terms that referred to the same thing: 'terrestrial TV' and 'airwave TV', but these days we only see 'terrestrial TV' possibly because the government issued a unified position.
3. The DTTV and the analogue television services must be in operation simultaneously during the transitional period.
4. Restrictions are placed on the transmission and its direction until the problem of radio interference with the existing analogue service, which is discussed below, is resolved.
5. In the white paper, 'Information and Communications in Japan' (Ministry of Internal Affairs and Communications, 2004), feature 2 is split into two independent features instead of indicating feature 3 to stress the superiority of the service.
6. A long-distance relay race between Tokyo and Hakone in which universities compete. It is held on 2–3 January every year and has a history of 80 years.
7. A popular programme that has been aired at the end of every year for more than half a century.
8. Neither the Radio Law nor the Broadcasting Law includes the term, 'universal service'. It is a concept borrowed from the telecommunication industry (Hayashi and Tagawa, 1994). However, the Ministry of Internal Affairs and Communications (MIC) has required terrestrial television broadcasters to ensure the universality of their terrestrial television services. Specifically, this administrative guidance has been given directly pursuant to Article 7 of the Broadcasting Law to NHK, and in the form of the Basic Broadcasting Plan (prescribed in Article 2-2 of the Broadcasting Law) to commercial broadcasters.

9. Both the Radio Law (paragraph 4 of Article 5) and the Broadcasting Law (paragraph 1 of Article 2) define broadcasting as 'the transmission of radio communication intended to be received directly by the general public'.

10. The governmental organisation reform in 2000 integrated the former Ministry of Posts and Telecommunications (MPT) with the former Ministry of Home Affairs and others into the Ministry of Internal Affairs and Communications (MIC).

11. All the data are based on the number of households. Japan has approximately 48 million households while there are approximately 100 million TV receivers. It means that every household owns slightly more than two such devices on average.

12. In the broadcasting industry, this scheme is called 'the separation between hardware and software'. It is easy to understand if you think of layer-by-layer separation that has become a common practice in the computer industry (like the division and collaboration between Intel, specialising in chips, and Microsoft, specialising in operating systems) (Hayashi and Ikeda, 2002).

13. The concept itself had been presented at ITU Telecom 79 in Geneva in 1979. But this is the first time it was published in the English-language literature that is available so far.

14. The following analysis is based mainly on the observations of telecommunications networks, which have an interactive nature. However, the developments in economic analysis since 1985 have discovered the similar phenomena between hardware (conduit) and software (content). And it is now considered to be general theory applicable to any networks.

15. Before the mobile phone revolution, Negroponte (1995) predicted the 'Negroponte revolution', which refers to the shift of broadcasting to wired connections and the shift of communications to wireless connections.

16. The number of TV units installed, which is approximately 100 million, is merely based on shipments. This figure does not indicate the number of units in operation. The number of units installed at home is accurately determined as this information is necessary for NHK in collecting subscription fees. But there is no statistical data on those installed in offices. It should be compared with a situation in which CNN and Bloomberg are watched by people in offices.

17. A specific example is that it is detrimental to users that NHK's subsidiaries are subject to the limitations on activities and that NHK's Internet services are restricted. This also reduces Japan's national competitiveness. (Hayashi, 2001)

18. Everybody in the industry has together expressed their opposition to the proposal of the IT Strategic Headquarters on horizontal separation, that is, the separation of hardware and software in the broadcasting industry. It is inevitably understood that it is because they are in such a special industrial environment that they lack the common sense of society. The structure of the regulations is one thing and corporate management is another. Horizontal separation of the regulations does not necessitate separation of corporate organisations in the same manner. If the corporate structure needed to conform to the legal structure, then, under the current legal system, broadcasters would have to be split into companies controlled by the Radio Law and companies controlled by the Broadcasting Law.

19. Universality could be attainable by means of direct reception from satellites or retransmission on cable television services, but Japan has persisted in terrestrial broadcasting. Consequently, cable television is positioned as a mere supplementary medium. Multisystem operators (MSOs) had been prevented from development of the industry until the mid-1980s.

20. This fact may coincide with the conclusion in Mitomo–Ueda paper (Chapter 13).

21. Here the author analyses the issue as far as it relates to digitisation. For more general discussion especially in relation to content, see Nakamura (2004).

22. The provisions of the Broadcasting Law are consulted in regulating the mass media as a whole because there is no law that regulates newspapers and publishers. However, it is necessary to pay attention to the fact that, with regard to newspapers, there is a provision in Article 1 of the Daily Newspaper Law that stipulates that 'the articles of

incorporation may limit the transferee of shares to those involved with business of the joint-stock corporation' and that it constitutes an exception to the Commercial Code.
23.   This ministerial ordinance is as powerful as an ordinary law because it originates from a rule set forth by the 'Radio Regulatory Commission', which was an independent administrative commission in 1952.
24.   There is an argument that this rule infringes the First Amendment of the United States' Constitution, given that it deprives cable television operators of their freedom of programming. Its constitutionality was actually tested in a lawsuit. Today, the Cable Television Consumer Protection and Competition Act of 1992 provides for the obligation to carry local programmes, but the debate is continuing.
25.   Yutaka Shigenobu is the president of TVMAN UNION, Inc., the most highly reputed TV production company in Japan. Hirokazu Kore-eda, who directed the movie that won the award for Best Actor at the Cannes Film Festival in 2004 entitled 'Nobody Knows,' is an employee of this production company.

REFERENCES

Amaya, Ryuji (2004) 'Beginning of digital terrestrial television and end of analogue system' *InfoCom Review*, **33** (in Japanese).
Fujitake, Akira (2000) *Japan's Mass Media Illustrated*, NHK Publishing (in Japanese).
Funada, Masayuki (2002) 'A proposal to revise the media concentration rule', *Rikkyo Law Review*, November (in Japanese).
Hayashi, Koichiro (1992) 'From Network Externalities to Interconnection' in Cristiano Antonelli (ed.) *Information Networks*, Amsterdam: North-Holland.
Hayashi, Koichiro (1998) *Networking – Economics in the Information Age*, NTT Publishing (in Japanese).
Hayashi, Koichiro (1999) 'Wake-up broadcasters! – Seven myths on digital Terrestrial Broadcasting', InfoCom Research (ed.) *InfoCom Outlook 1999* NTT Publishing (in Japanese).
Hayashi, Koichiro (2000) 'Broadcasters! Be an ordinary industry' InfoCom Research (ed.) *InfoCom Outlook 2000*, NTT Publishing (in Japanese).
Hayashi, Koichiro (2001) 'A comment on the scope of business by the subsidiary companies of NHK' submitted to MPT (in Japanese) http://lab.iisec.ac.jp/~hayashi/comments.html.
Hayashi, Koichiro (2002) *Information and Communications Industry*, ICIEE (Institute of Communication, Information and Electronic Engineers) (in Japanese).
Hayashi, Koichiro and Yoshihiro Tagawa (1994) *Universal Service*, Chuo-Koron-Sha (in Japanese)
Hayashi, Koichiro and Nobuo Ikeda (eds) (2002) *System Design for the Broadband Era*, Toyo-Keizai-Shimposha (in Japanese).
Hirai, Takuya (2003a) 'Rethink DTTV' (in Japanese) http://www.hirataku.com/seisaku/seisakus/ti-diji-teigen.htm
Hirai, Takuya (2003b) 'Revise DTTV plan for its success' *New Media*, August (in Japanese).
Ikeda, Nobuo (2003) 'Spectrum buyouts: a mechanism to open spectrum' (revised December 2003), http://www.rieti.go.jp/jp/publications/summary/02030001.html.

Kitagawa, Masayasu (2003) 'Broadcasters should publish their own manifesto for DTTV' *New Media*, December (in Japanese).

Kitahara, Yasusada (1984) 'INS (Information Network System) – telecommunications for the advanced information society' *Computer Networks 8*.

Ministry of Internal Affairs and Communications (2004) *InfoCom White Paper*, Government Printing Office (in Japanese).

NAB (National Association of Commercial Broadcasters) (1998) 'Management of commercial broadcasting in the digital age', NAB Report (in Japanese).

NAB (National Association of Commercial Broadcasters) (1999) 'Management of commercial broadcasting in the digital age (revised version)', NAB Report (in Japanese).

Nakamura, Kiyoshi (2004) 'Broadcasting Industry' in Hideki Ide (ed.) *Regulation and Competition in the Network Industries*, Keiso Shobo (in Japanese).

Negroponte, N. P. (1995) *Being Digital*, Cambridge, MA: MIT Press.

Noam, Eli M. (1992) 'A theory for the instability of public telecommunications systems' in Cristiano Antonelli (ed.) *Information Networks*, Amsterdam: North-Holland.

Noam, Eli, Jo Groebel and Darcy Gerberg (eds) (2004) *Internet Television*, Mahwah, NJ: Lawrence Erlbaum Associates, Publishers.

Research Institute of NAB (2000) *Future of Digital Broadcasting*, Toyo-Keizai-Shimposha (in Japanese).

Rogers, Everett M. (1962) *Diffusion of Innovations*, New York: Free Press.

Shigenobu, Yutaka (2004) 'A culture model theory for DTTV' *New Media* May (in Japanese).

9. The management of digital rights in pay TV

Campbell Cowie and Sandeep Kapur

INTRODUCTION

Anyone who legitimately consumes an information good[1] or a digital media product, say, by viewing a DVD, by watching a sporting event on a pay per view basis or by downloading a music track to their i-Pod device, also consumes in parallel some form of content protection. The form of content protection chosen by the content creator or distributor will typically depend on the nature and value of the underlying content. Content suppliers with low value content and few piracy concerns may rely on legal protection under copyright laws. Suppliers with more valuable content may supplement legal protection with technological solutions. Consider, for instance, the standard model of conditional access in pay TV which uses the encryption of television signals to restrict access of programming to paying customers. In general there is heterogeneity of demand for content protection which results in the use of a wide range of non-mutually exclusive legal and technical solutions.

As the growth of broadband Internet increases the potential for illegal access to, and exploitation of, content, and as the commercial value of that content increases, we expect a change in the balance of legal and technological solutions to protect intellectual property. Even as a case is made for adapting copyright laws to the new technological landscape,[2] more sophisticated technological means of content protection are being developed. We believe that one such technological solution, Digital Rights Management (DRM), will become increasingly important for protecting digital content in the pay TV supply chain and in time is likely to supersede conditional access. DRM describes a suite of software and hardware technologies that can be deployed to provide persistent end-to-end content protection.

Our chapter aims to contribute to the understanding of the potential impact of DRM on market power in digital pay TV. As we aim to identify long term trends in a technologically dynamic area, our analysis can only be speculative. Following this introduction we provide some non-technical background on what DRM means and, using analogies from related markets, suggest new business models that may be feasible in the pay TV environment. In the third section, we explain the roots of DRM development, which can be traced to measures taken to counter the illegal theft of intellectual property and many of the concerns about the potential that DRM has to strip consumers of their usage rights arise. Following this, in the fourth section, we offer suggestions as to how we believe DRM may impact on the markets for intermediary services in the supply chain and in the fifth section we explore the potential impact of DRM on selected competition policy issues common to pay TV markets. The final section concludes.

BACKGROUND TO DIGITAL RIGHTS MANAGEMENT

Digital goods are protected through a variety of methods. Copyright laws provide legal protection, but these require that the rights owner is alert to any breaches of those legal protections and is in a position to enforce those rights. Legal protection may be supplemented with simple technical barriers to easy duplication of content. For example, the deliberate use of faint type in printed documents makes it harder to produce perfect photocopies. As the commercial value of the digital content increases, and as the growth of the Internet increases the ease of illegal access, storage and distribution, we should expect that more sophisticated forms of protection will arise. Even as there is talk of the potential of quantum encryption (which involves utilising the principles of quantum mechanics to generate random light-based keys, rather than large numbers, to secure content),[3] the content creation and delivery industries are concentrating on the shorter term development of an efficient system of DRM to provide end-to-end persistent technical protection for their content.

DRM has been described as 'a systematic approach to copyright protection for digital media',[4] though this does not quite explain what DRM is, or how it might evolve in the future. From a technical perspective DRM encompasses 'the description, identification, trading, protecting, monitoring and tracking of all forms of usages over both tangible and intangible assets'.[5] More simply, DRM is a suite of solutions that allow for the technical identification and protection of intellectual property. The technical protection allows the owner to monitor and control access to the protected content from the point of its creation to the end of its life. Usage rules embedded in the content 'metadata' can control how and by whom a particular piece of content can be exploited. This control can include the designation of devices upon which the content can be consumed and, whether or not it can be transferred from one platform to another. It can, in theory, allow for differential charging for access based on the identity of the user, time of use or the device on which it is used. This empowerment of content owners gives rise to a broader understanding of DRM, which encompasses how intellectual property rights are managed and exploited in a digital environment.

DRM technologies employ a number of discrete tools, which are combined in order to provide the kinds of functionality described above. The primary tools are briefly described in Table 9.1.

The precise choice of tools in any DRM solution will depend on the purpose of the DRM and the nature of the content that is to be protected. Tools used to protect documents differ from those used to protect a Hollywood movie. For instance, Adobe Acrobat is software that utilises secure containers to set permissions on how protected documents can be used and distributed. Copy Generation Management System (CGMS) comprises rights expression languages that control the ability to duplicate DVDs, as do copy management rules for DAT and minidisk players which are managed by a Serial Copy Management System (SCMS). Similar tools underpin the Content Scramble System (CSS) in DVDs, which allows content owners to segregate markets geographically, by restricting usage of regionally-encoded DVDs to hardware specific to the regions.

In addition to the threat posed by physical piracy, the growth of broadband has led to a growing online market for content, which requires appropriate DRM technologies to provide rights protection. For example, Apple's successful online music retail service, i-Tunes, and AOL's Musicstore service have extensive DRM, exploiting user ID, authentication and billing tools. It is a form of DRM that prevents users of games consoles

Table 9.1   Some common DRM tools[6]

Tool	Purpose
Secure containers	Restricts access to those with authorisation
Rights expression languages	Establishes which users have access
Content description	Unique description of content for search purposes
User ID	Allows for the tracking of usage
Authentication	Determines an individual's usage rights
Fingerprinting/Watermarking	Complementary tools allowing the originator to identify unauthorised use
Payment	Billing and payment mechanism

from also using them as conventional PCs. Tools such as watermarking and fingerprints are increasingly used by rights holders to trace illegal copies of content or to degrade the quality of the copied material.[7]

The ability of the content owner to attach usage instructions to the metadata sent with the content enables greater control over the exploitation of the content. The implications are obvious. For example, while it may be technically feasible in the future for video content to be transferred from a Digital Video Recorder to a portable device, such duplication would deprive content owners from appropriating the value of the transfer to the user and possibly the potential value from the onward distribution of its content. This lack of appropriability is an obstacle to the development of business models based on transferable content and creates a clamour for legal impediments on technologies that enable transfers. DRM technologies can enable appropriability and provide the content owner with greater incentives to support new technological platforms.

DRM also offers to reduce the transactions costs associated with the transfer of digital media. These typically include the costs of search, the costs of reaching a contractual agreement, monitoring exploitation of the acquired good and enforcing contractual terms, and together may be prohibitively large in some transactions. As transactions costs fall new business models become viable.

The welfare benefits arising from the introduction of new products and the viability of new forms of exchange may be substantial. For instance, in the case of the US telecommunications market, Hausman (1997) estimates that the welfare costs of Federal Communication Commission's failure to introduce two new telecom products was approximately $2 billion.[8] Amazon.com provides an example of the welfare impact of the emergence of a transformative entrant rather than a new product. Brynjolfsson, *et al.* (2003) estimate that the increase in consumer welfare resulting from the introduction of online sales of books was between $731m and $1.03bn for the year 2000 alone. More than the effect of competition on retail prices, the gains in welfare came from the reduced cost to the consumer of searching and locating previously hard to find titles, and access to recommendations based on their identified preferences.

In much the same way as Amazon.com, the transformative power of DRM, and its ability to support new pay TV business models for the exploitation of existing content, may lead to substantial welfare gains. For example, a Video-on-Demand service with time-restricted usage could be a substitute for video rental, while a download service with unrestricted time usage could be an alternative to DVD sales. The ability of a service provider to collate information on consumer preferences by monitoring usage could enable the retail service to make valuable recommendations about alternative content that the consumer may be interested in – potentially generating welfare gains in much the same way as is done by Amazon.com.

PIRACY

The early development of many DRM technologies has been driven by growing concerns about piracy. The ease of illegal access to intellectual property via the Internet has become a major issue for both policy makers and content creators. The inability of existing copyright legislation to prevent the mass downloading, distribution and physical copying of content has created the need for improved technological solutions. However, there is a growing fear that increased reliance on DRM technology to combat piracy compromises the principle of fair use under conventional copyright law and may ultimately act to the detriment of society in general (see Samuelson (2003) and Lessig (2004) for good illustrations of the arguments).

Many characteristics of digital media products make them particularly susceptible to illegal piracy?

- Digital content goods are non-rival in consumption.
- Without recourse to technological solutions, digital content is non-excludable.
- The incremental cost of making a copy is insignificant relative to the fixed costs of making the original.
- The cost of transporting and storing digital media products is negligible and declining with falling PC prices and Internet charges.
- The cost of detecting piracy and tracing illegal use over the Internet is often high relative to the commercial value of the content.

The cost structure, specifically the near-zero marginal cost of making copies, means that cost-based pricing is unlikely to recover the fixed costs of producing the first copy. In such environments, it may be rational to charge customers according to their individual willingness to pay. Prices would then differ across consumers. Not only is this discriminatory pricing commercially rational, in many cases it is also Pareto efficient. The sequential release of content through time-specific windows – say, the cinematic release of a new movie, followed by its release on pay TV, followed by release on DVD, and so on, is a common form of inter-temporal price discrimination. This sequential release strategy, when supported by copyright law, grants the content creator monopoly rights over the exploitation of the content within each release window.

While this enables the content owner to set prices above marginal cost within each period, such pricing generates opportunities for profitable piracy. Illegal copying is hardly a new phenomenon, but digitalisation has made it possible to make near perfect quality copies at low cost. The Internet has enabled the low cost distribution of pirated content. The decentralised structure of the Internet makes it difficult to track usage of that pirated content, and since piracy often straddles many jurisdictions, it is costly to enforce legal action. The content creating industries face the risk that if pirated copies become a reasonable substitute for the legitimate version, many consumers will be reluctant to pay prices significantly above marginal cost. Rob and Waldfogel (2004) conducted a small-scale experiment to illustrate this: a sample of students reduced their expenditure on legitimate CDs from $126 per capita to $100 when offered the possibility of free downloads. If the results of this small scale experiment can be legitimately extrapolated, the production system underpinned by exclusion and price discrimination will no longer be sustainable. In the context of music file sharing this fear has been expressed by the Committee on Intellectual Property Rights and the Emerging Information Infrastructure (2000): 'For publishers and authors, the question is, how many copies of the work will be sold (or licensed) if networks make possible planet-wide access? Their nightmare is that the number is one.'

Identifying the cost to the content industries of illegal piracy of its intellectual property is not a simple task. At one level the download of an illegal film is a lost legitimate sale, so that the cost of the estimated 600 000 movies illegally downloaded each day in the US could be valued in terms of lost revenue at the box office or in terms of lost DVD sales (Deloitte, 2004). However the analysis is not so simple. Not everyone who purchases or downloads an illegal copy would have bought a full-priced legitimate copy, so that valuing every lost sale at full price overstates the loss. Additionally, not all sales might have been lost. Sometimes consumers sample music by downloading a track illegally, and subsequently purchase a legitimate copy of the album. Hence it is not easy to quantify the cost of piracy to content owners.[9]

While piracy erodes the revenue of content owners, measurement of the impact of piracy on overall economic welfare is not easy. Against the loss of producer welfare, there is the possibility of an offsetting increase in consumer welfare. Piracy lowers average prices, by giving consumers the option to switch to a cheaper (although possibly inferior) version of the product. Piracy also extends the market: some consumers who were unable or unwilling to buy products at the full retail price can buy the cheaper (albeit illegal) pirated version.

And while piracy usually hurts producers, sometimes the loss may be partially or substantially mitigated through indirect appropriation of the value of pirated content. Besen (1987) provides an interesting example. A television channel may be able to charge higher fees for advertising if it can show that its content is being pirated: the additional viewer base may increase the attractiveness of the channel to advertisers. It may even be that the piracy-driven increases in advertising revenue can more than compensate for the loss in carriage or subscription fees from piracy. Liebowitz (1985, 2002) argues that publishers of periodicals realise that library copies of their content may be vulnerable to piracy through photocopying, but may be able to extract some surplus through higher charges for institutional subscriptions than for individual subscriptions. The libraries themselves may recover some of this through profitable pricing of photocopying. Similarly, the application of a levy on blank recordable media like tapes, CDs and DVDs may be viewed as a means of indirect appropriation. In these scenarios the content creator can extract some of the surplus from the act of piracy, provided that the pirate can be identified and a mechanism for transfer can be imposed. If so, piracy underpins a form of price discrimination, and producer surplus may be higher than what is achievable in the absence of piracy. Although nice in theory, the inability to identify the pirate, establish the value of the copies that the pirate will produce and enforce the higher price makes indirect appropriation impractical in most cases (Liebowitz, 2000).

The introduction of dynamic considerations provides for additional complications. Even when piracy of software results in lost revenue to producers, there may be offsetting benefits. This may happen if, for instance, there is an element of lock-in to a particular piece of software or a network effect surrounding the adoption of a particular standard. Students who use pirated copies of software are more likely to pay for newer versions of that software (as well as complementary products) later in life, if only because they are more familiar with its routines. Moreover, as future generations of the software are developed and marketed, the illegal copies of the earlier generation may serve to lock consumers into legitimate purchases of later generations. Similarly, pirated copies of an earlier (later) instalment of a part-work, for example *Star Wars* or *Lord of the Rings* could serve to increase sales of legitimate copies of later (earlier) episodes. Ultimately, the net effects of piracy are context specific.

Traditionally, content owners have relied on a mix of technical and non-technical means to minimise the adverse effects of piracy, including:

- use of copyright law to deter piracy;
- raising the costs of piracy, by making it expensive to replicate quality (Novos and Waldman, 1987);
- bundling the IP asset with a complementary product that cannot easily be copied (Novos and Waldman, 1987 and Besen, 1987);

• lowering the potential return to piracy by offering discounted versions to compete with pirates on price and manipulation of the release window strategy.

Going forward, technical solutions to piracy are likely to come to the fore. DRM is considered a key technical component for anti-piracy efforts, but as with any software it is susceptible to hacking. This is recognised in the drafting of the EC Copyright Directive (2001), which prohibits measures designed to circumvent technical protections measures such as DRM.[10] This is crucial for rights owners. Not only is the intellectual property protected through the power of the courts, but the technical measures deployed by rights owners to protect the intellectual property are also protected.

The scope of protection provided to content owners through DRM differs from that provided by copyright law. While copyright laws accord a degree of monopoly control over content, exceptions to the law are normally incorporated to remedy the potential for abuse of this monopoly award and to protect social interests more generally. The doctrine of first sale (also known as the principle of exhaustion) serves to limit the exclusive right to distribution of the content. A consumer who purchases a legitimate item is entitled to sell, rent (in certain circumstances only), donate or share that item with others. In so doing, the market power of the copyright owner is constrained by what amounts to a second hand market. Second, what is generally known as the fair use doctrine enables certain classes of users to exploit the copyright protected content without the owner's consent, where the specified usage is deemed to be in the public interest. Although the doctrine is peculiar to US copyright law, it is mirrored in the laws of other territories.

In contrast, DRM is a technical means of exclusion and requires explicit action from copyright owners before consumers can take advantage of the traditional exceptions provided for in law. DRM allows the owner to specify usage rules for each piece of protected content, restricting exploitation to a specified set of uses, whether it be read only, store for a certain time, edit, distribute a specified number of times or view only on specific devices.

Indeed there are some who believe that the roll-out of DRM technologies grants copyright owners too much control over how their content is exploited and consumed (Samuleson, 2003). These concerns are not new.[11] However, as discussed above it is not clear that complete exclusion is always in the interests of copyright owners: limited piracy may be consistent with commercial advantage. As Varian (1998) notes, the objective of content owners is to maximise the value of their content, not to maximise the protection. It is not therefore certain that the fears of many consumer lobbyists are justified.[12]

While the European Commission has aimed to providing a comprehensive framework of protection for technical measures such as DRM, consumer interest groups have continued to challenge the scope and nature of the protection. As Table 9.2 shows, the legal situation remains uncertain.

Looking ahead, content creators are increasingly likely to support the use of DRM technologies as a means of content protection. The greater technical ability to protect content through DRM technologies may also enable content creators to manage more sophisticated windowing strategies, to introduce pricing structures that are more flexible and tailored to individual consumers' demand. Flexible pricing and price discrimination will allow content owners to offer prices that more closely reflect individual consumers' demand than is feasible today. This may support pricing to

Table 9.2   DRM and the law: recent ambiguities

Country	Issue summary
UK	The High Court held that modification chips allowing users to defeat DRM protections on Sony video game cartridges (which specify regional coding) were in violation of the Copyright Directive.[a]
Spain	A court upheld the consumers' right to hack the same regional coding protections.[b]
France	The courts have forced EMI to withdraw copy-protected CDs from shops in response to complaints that the copy protection DRM prevented consumers from exercising the right to make private copies (Tribunal de Grande Instance de Nanterre, 2003). However, in another case the court refused to uphold a challenge to the right of a film studio to incorporate DRM protections on the DVD of *Mulholland Drive* (France Tribunal Paris, 2004)
Norway	In perhaps the most famous case, courts upheld the right of consumers to hack CSS, the DRM protection of DVDs, using DeCSS[c]
Belgium	A court ruled in May 2004 that consumers had no right to make personal copies of copyright protected material so had no right to hack DRM protections (Tribunal Bruxelles, 2004)

Notes:
a.   See http://news.zdnet.co.uk/hardware/emergingtech/0,39020357,39161307,00.htm
b.   See http://www.geek.com/news/geeknews/2004Jul/bga20040723026148.htm
c.   See http://www.linux-magazine.com/issue/28/WorldNews.pdf.

certain consumers that reduces the margin between the legitimate and the pirated product. The increase in output that may also be facilitated may go some way to addressing concerns of consumer groups about access rights.

INTERMEDIARIES: MANAGEMENT OF DIGITAL RIGHTS

The promise of security and control over how content is exploited across the supply chain is likely to have a significant impact on how content owners engage with consumers. While the public policy debate has focused on the issues of intellectual property and piracy in the recent past, the emergence of new business models for delivery of content will shift the policy debate towards regulatory and competition policy challenges arising from the new technologies. In particular the consequences of the new developments for the intermediaries in the supply chain will come under considerable focus.

Content creators vary in their approach to the distribution and retailing of their content. While some control the process end-to-end (these may be viewed as conventional business-to-consumer operations), others merely license their content to others for distribution and retail (and thus are business-to-business operations). Traditional, free-to-air producers/ broadcasters, such as the UK's BBC and ITV, are representative of the former model, while independent producers and film studios sit within the latter group. The different market models may reflect differences in the value of the content being distributed relative to the benefits in each case of using intermediaries to distribute the content.[13] Where the transactions costs involved in using intermediaries are high relative to the value of the content then the end-to-end model will typically be favoured.

Amazon and iTunes provide helpful illustrations of the role to be played by intermediaries in digital markets. They exist as a result of imperfections in the market for the delivery of digital products and services, but have been able to establish a presence in the value chain as a result of entry barriers falling due to technology and DRM. Although they operate in the Internet space, they may provide a useful precursor of the role of intermediaries in the broader digital chain and for pay TV in particular.

Bailey and Bakos (1997) identify the four major functions of an intermediary as:

1.   the aggregation of supply and demand;
2.   providing trust to transacting parties;
3.   market making; and
4.   matching buyers and sellers.

In the context of pay TV, the typical vertically integrated platform operator can be seen to be offering bundles of those functions. Their intermediation services include the retailing of channel bundles (aggregation of
supply and demand), the supply of conditional access and subscriber management (provision of trust) and the provision of electronic programme
guides (aggregation and matching).

With the increasing penetration of both digital TV and broadband
Internet, the recent interest in the economics of the digital supply chain is
not surprising. In many of the supply chains studied in the recent literature
the intermediary provides a matching role, putting consumers in touch with
suppliers.

Bhargava and Choudhary (2004) explain the network effects at work in
the supply of aggregation services by 'infomediaries'.[14] They use the
example of Expedia.com, which allows consumers to browse and book
holidays from a selection of suppliers, to explain how the number of buyers
using a service will increase as the number of suppliers increases. As they
note, the 'intensity of the aggregation benefit provided by an intermediary
to buyers (sellers) is determined by market characteristics and the mix of
information processing features made available to buyers (sellers)'. A
similar network effect exists in the provision of electronic programme
guides in pay TV and in the bundling of pay TV channels into retail packages. In pay TV the platform operators will typically not provide a matching role, but will act as a wholesaler, in effect managing the transaction with
the consumer on behalf of the content owner.

The role played by platform operators in providing intermediate services
will come under pressure as a result of the roll-out of DRM solutions in the
pay TV supply chain. This pressure may be increased as the ability to deliver
Internet-protocol based TV services over broadband lowers entry barriers
at the intermediate levels. In such an environment, the aggregation benefits
could conceivably be supplied by the technology as well as by new entrants
such as search engines, rather than by the existing players. However, for the
purposes of this chapter we concentrate on the impact on the intermediate
markets of a successful roll-out of DRM solutions. Table 9.3 summarises
how DRM solutions can substitute for functions that are currently provided by the platform operators.

In effect, DRM may enable content creators to do without some of the
functions currently supplied by intermediaries. However, this potential dis-
intermediation does not necessarily mean the content creators can do
without intermediaries altogether. In the pay TV chain, the intermediaries
shoulder the risks involved in dealing with end customers. Even in the pay-
per-view window, perhaps the TV window where the consumers is 'closest'
to the content creator, the platform operators manage the risks involved in

Table 9.3.   DRM substitution for intermediary functions

Role played by intermediary	In what way is DRM a substitute?
Provide transactional trust for consumer and producer	DRM may be used to unlock content only after payment and to extract payment only when the content has been supplied
Record transactional data	DRM can allow the content owner to monitor content exploitation
Protection against IP theft	Access can be restricted only to those who pay
Setting boundaries on fair use	DRM provides a technical replacement for legal boundaries on fair use exploitation

both retailing the product to consumers (subscription management, bill processing and marketing) and in licensing the content (estimating effective demand). For this, intermediaries such as BSkyB are well rewarded.[15] Where DRM can fulfil some functions of intermediaries – in particular the authentication of users/subscribers and the management of transactions across platforms – there is the potential for content owners to integrate down the supply chain to the relevant stages. However, the scope for the complete displacement of intermediaries is limited:

- *Brand awareness*. With the exception perhaps of Disney there are few content companies with an established brand that consumers can identify with a particular genre of content. Content creators would have to invest heavily in brand awareness.
- *Business model*. As most content creators operate as business-to-business entities, downstream integration would require radical restructuring of their operations, including investment in consumer management at the retail level.
- *Aggregation*. Although advanced search techniques may reduce the value-added to consumers of aggregation undertaken by platform operators, the platform operator does provide scale and scope economies in the supply of content.

In short, digitalisation, underpinned by DRM, will not automatically lead to the elimination of intermediaries in the pay TV supply chain. We are more likely to see a change in the economics of the supply chain at the intermediate levels. Some functions currently provided by the traditional intermediaries may no longer be necessary in the delivery of many of the new services. In areas such as video on demand (and its variants) there may be less need to package content into channels as consumers will be able to search for content on a title by title basis or by genre. The lowering of entry barriers in the intermediate markets, as a result of the potential for unbundling of services, means that there is scope for independent niche intermediaries, to compete with established players who will continue to provide bundled services. For an illustration of what the future market structure could look like, consider Movielink, a joint venture between a number of leading Hollywood film studios.[16] Movielink offers broadband consumers the opportunity to download movies to their home PC or TV, either for immediate view or for storage. Search costs are low as consumers can select directly from a wide range offered by the studios involved in the joint venture. In effect, Movielink has enabled the studios to bypass the established intermediaries in a particular segment of the home movie viewing market. As the new intermediaries at the aggregation stage can expect to get regulatory support for gaining access to electronic programme guides, consumer search costs need not increase much as new players arrive in this market. Similarly, where content can be grouped by genre, new players can easily package their content in a well-defined offering.

Where the genre is hard to define, leading to higher search costs, or where the transactions costs involved in managing multiple intermediaries are large relative to the value of the content, the traditional intermediaries are likely to retain some advantage in packaging content into discrete channels. Here content owners would continue to benefit from the bundled service provided by intermediaries. Even here, the potential competition from the new entrants is likely to provide content companies with greater leverage with intermediaries.

Where content is valued across multiple platforms, there may be greater scope for content creators to assume some intermediation functions using DRM. For example, consumers may wish to download a movie on to a DVR and then transfer that to other devices. In such contexts it may be simpler to use multi-platform DRM enabling a single user authentication and billing system than transacting with multiple, platform-specific intermediaries.

IMPLICATIONS FOR REGULATORY AND COMPETITION POLICY

Successful roll-out of DRM solutions in the pay TV supply chain is likely to transform the economics of pay TV. If so, it would require a change to the traditional case for regulatory intervention to constrain the potential for the abuse of market power in this market. We focus on two underlying features of DRM in this context.[17] First, DRM provides a channel for secure, delivery of content through open distribution networks, and may allow content owners to bypass the existing pay TV channels built around proprietary platforms. More generally, the creation of an alternative delivery channel will alter the distribution of market power in the supply chain. Of course, the outcome will also depend on the market structure that emerges in the DRM technologies sector. Second, the increasing use of DRM in the delivery of content will have significant implications for the nature of pricing in these markets by altering the scope for price discrimination. We look at each of these issues in turn.

DRM and Market Power in Pay TV

Figure 9.1 provides a simple illustration of a digital pay TV supply chain. The key regulatory bottlenecks in the current supply chain are well known and have been subject to regulatory investigation (Competition Commission, 1999 and OFT, 2002) and to academic scrutiny (Cowie and Williams, 1997). The market power of the platform operators is allegedly built around their exclusive access to valuable content through long-term contracts for premium sport and blockbuster movies. Vertical integration across proprietary conditional access services has also generated regulatory concern. Despite several attempts to moderate the market power of the platform operators, using both *ex ante* regulation (of conditional access) and *ex post* competition law (in cases relating to access to and the licensing of sports and movie rights), the market power of large platform operators remains virtually intact, with little sign of effective potential entry.

What are the sources of market power of the current pay TV platform operators? Pay TV operation requires considerable fixed investment to set up the technical infrastructure for programme delivery. Further, investment in editorial infrastructure is necessary to aggregate licensed content into channels and to develop own content. The first category of expenditure involves economies of scale while the latter involves economies of scope. Such a cost structure – and the fact that many of the costs are in the nature of sunk costs – creates a tendency towards concentration. Consider the relatively concentrated nature of most pay TV markets in Europe. Not surprisingly, even in countries with some competition between platform operators, there is considerable commercial pressure for consolidation.[18] The scope for competitive entry in this sector is limited: entry usually requires an operator-specific decoding device (the set-top box), which results in significant consumer switching costs and some element of lock-in.

The diffusion of DRM solutions may challenge the market power of pay TV operators, and even ease some of the regulatory and competition policy concerns. DRM can enable the secure and exclusive delivery of content

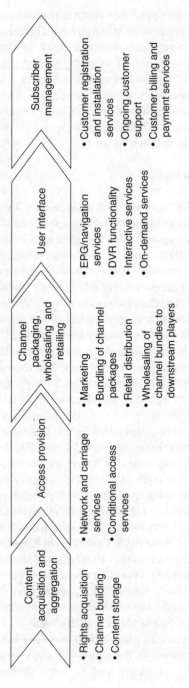

Figure 9.1 Illustration of a hypothetical digital pay TV supply chain

Content acquisition and aggregation
- Rights acquisition
- Channel building
- Content storage

Access provision
- Network and carriage services
- Conditional access services

Channel packaging, wholesaling and retailing
- Marketing
- Bundling of channel packages
- Retail distribution
- Wholesaling of channel bundles to downstream players

User interface
- EPG/navigation services
- DVR functionality
- Interactive services
- On-demand services

Subscriber management
- Customer registration and installation services
- Ongoing customer support
- Customer billing and payment services

through open networks, and thereby obviate the need for proprietary networks built around encryption-based hardware. If DRM allows content owners to bypass conditional access, it is likely to weaken the market power of the current vertically integrated platform operators and strengthen the bargaining position of the upstream content owners/creators in the wholesale market for their content and channels.

To understand this, consider the market for the licensing of content rights for pay TV. Due to the relatively small number of buyers and sellers in this market, it is not a competitive market in the textbook sense.[19] If there are multiple pay TV operators vying for a licence, the content owner may choose to auction the rights. If there is only one pay TV operator that can credibly acquire the rights from a content owner – that is, if there is a bilateral monopoly – the terms of the licence will be determined through negotiation between the licensor and licensee. While each party may prefer an outcome that maximises its share of the profit from the commerical exploitation of content, the outcome of the bargaining process would depend naturally on their relative bargaining strengths. For instance, if their bargaining strengths are equivalent, the agreed upon price will be mid-way between their individual positions in the event of disagreement. Of course, the outcome may be sensitive to the availability of outside options. For each player, the outside option refers to what players expect to get if they fall back on alternative opportunities in the event of disagreement in the bargaining process. In general, the availability of more valuable outside options is likely to improve a player's payoff (and certainly cannot lower it) within the bargaining process.[20]

A content owner is naturally attracted to the dominant pay TV platform/ channel. The dominant channel may have considerable advantage over its rivals due to its larger subscriber base, its superior knowledge of down-stream demand or its ability to package and promote the content better than other channels. The dominant operator can exploit these advantages and the related advertising potential to maximise the surplus to be shared between the licensor and licensee. A content owner's outside options may be quite limited if rival pay TV operators have a small subscriber base and if there are no other channels for distribution of content. In contrast, from the dominant channel/platform operator's perspective there are few forms of content for which there are no credible substitutes. Typically they have stronger outside options. This asymmetry of outside options implies that the bargaining outcome is likely to be favourable to the pay TV operator. It is in these circumstances that competition authorities, content owners and independent channels have raised concerns about the behaviour of the vertically integrated platform operators, alleging abuse of market power.

The roll out of DRM can potentially alter the bargaining outcome. Content owners can threaten to use DRM technologies to distribute the content directly to their consumers and, in the process, bypass the primary bottlenecks and intermediary services of the pay TV platforms. Of course new distribution channels will need to reach agreements with the vertically integrated platform operator to secure access to the programme guide and to other technical access services for effective delivery of content. Over time, there is scope for unbundling of such intermediary services, so that DRM-enabled delivery will become a credible alternative. In terms of bargaining theory, DRM increases the outside option for the content owners. Where the outside option is credibly more profitable than the existing outcome for content owners, the emergence of this better outside option should enable content owners to improve on their bargain relative to pay TV platform operators. This is not to say that content owners will necessarily exploit the new outside options. Rather the increased leverage that the new options provide can allow them to secure a greater share of the rent derived from marketing their content via existing platforms.

The recent history of auctions of television rights for Football Association Premier League (FAPL) matches suggests how the balance of bargaining power might shift. The FAPL has tried to suggest that it may launch its own channel if it cannot secure favourable terms from BSkyB,[21] but the threat has never been credible. It is undermined by the fact that a channel owned by the FAPL would have to negotiate carriage with BSkyB, as well as access to the intermediary services provided by BSkyB and its related companies, and BSkyB would be in a strong position to appropriate the rents at that stage. If so, FAPL may be in no better position than if they had agreed to license the rights to BSkyB in the first place. However if FAPL is able to license DRM solutions from a party other than BSkyB, the expected returns to FAPL from launching their own channel would be higher, and this enhanced outside option makes FAPL's threat to launch their own channel more credible. Similarly, film producers, particularly the major Hollywood studios, would benefit from having the outside option of retaining their own rights and developing their own distribution services as an alternative to licensing to a downstream channel operator. Indeed, Movielink is an early example of how this might pan out.

What are the implications of this for regulatory and competition policy? If the threat of DRM-enabled distribution channels is able to curtail the market power of the dominant pay TV platform operators, it would ease some of the current regulatory concerns. However, there are some caveats. The dominant position of pay TV operators may only partly be based on conditional access. It could also derive from their production and editorial functions, say those associated with the packaging of content into attractive channels that match subscriber preferences, or their reputation for innovative content production. While DRM can level the playing field in secure delivery of content, superiority in the editorial or production infrastructure will preserve some advantage for the current pay TV operators. In such circumstances, DRM will enable only limited displacement of the pay TV operators, more so for some categories of content than others. Regulatory policy may then focus on whether retention of these advantages requires regulatory oversight.

It is possible that the greater deployment of DRM will only shift market power from pay TV operators to the upstream content owners/creators, without affecting the overall degree of monopoly power. In many markets the current regulatory concern has been with the arrangements/market forms by which rights to content are transferred from content creators to pay TV broadcasters. If monopoly power migrates upstream, regulatory oversight will need to make an appropriate transition itself, to see if the growing dominance upstream is against the public interest.

Further, it is possible that some market power might shift to the technology vendors that came to control the dominant DRM standard(s). It is conceivable that slight market advantages – real or perceived – could cause the market for DRM applications to tip in favour of one or two proprietary standards. If so, one bottleneck technology (proprietary platforms of pay TV channels) would have been replaced by another (proprietary DRM technology), with no overall improvement from the consumers' point of view.

This scenario cannot be ruled out but is not a foregone conclusion. The outcome may well depend on many technological and strategic considerations. Entry in the DRM applications market might be easier than launching a pay TV platform, so that multiple DRM standards could potentially coexist in a competitive environment. Heterogeneity in patterns of demand for DRM may create scope for distinct categories of DRM applications. As different types of content are susceptible to differing degrees of piracy, content owners will seek different levels of protection. For some, current encryption technologies provide sufficient protection, while others, like blockbuster movies, might prefer persistent end-to-end protection. For enterprise DRM (DRM designed to protect documents when there is shared access within the enterprise), methods akin to using faint type to make photocopying difficult might suffice. It is plausible that the heterogeneity in demand will sustain a variety of DRM standards in the long run. If so, any sub-market within a spectrum of protection technologies might be contestable.

The degree of interoperability between rival DRM applications will also affect the effective competition in this market. There will be pressure from the content industry for many of the different DRM solutions to be interoperable. Re-versioning content for different distribution platforms increases the costs of distribution, so that greater interoperability in DRM systems will allow them to reach more consumers at any given cost. The incentives for DRM operators to build in interoperability are more compli-cated. For one, interoperability increases the intensity of price competition between rival standards, so that restricting the degree of interoperability might be attractive.

Overall while it is conceivable that the market for DRM standards may be concentrated, the heterogeneity of demand and the desire for intero-perability of reception devices from the perspective of content suppliers is likely to restrain this tendency. Competition policy may need to maximise the possibility of a competitive outcome, at least in terms of pricing of DRM technologies. Regulatory policy may also need to guard against the possibility that existing pay TV operators may themselves acquire interests in DRM/DRM vendors in order to prevent the dilution of their market power.

DRM, Pricing of Content and Consumer Welfare

Pricing structures for premium content, particularly movies, display con-siderable price discrimination. This is hardly surprising. Given the tech-nology of generating content – high fixed costs of producing the first copy, and near zero cost of subsequent copies – there is little reason for prices to be related to the marginal cost of production. Once that anchor is lost, pricing of content is likely to be related to consumers' willingness to pay and thus may well involve different prices for different consumers.

For price discrimination to work, a seller must be able to sort customers (that is, distinguish those who are willing to pay a high price from others who are not) and to segregate the market (so that those who are offered low prices cannot transfer the good or service to others facing higher prices). We argue that DRM technologies potentially could enhance content owners' ability to sort customers and to segregate markets, thereby increas-ing the scope and extent of price discrimination. DRM is likely to improve the precision of conventional price-discrimination strategies but also enable newer forms of price discrimination. Indeed it could be argued that the real impetus for DRM roll out comes not from concern about piracy, but from its potential for enhancing the value of intellectual property rights through more aggressive price discrimination.

A common strategy in the market for cultural content involves inter-temporal price discrimination (Caves, 2001). This relies on a simple attribute, impatience, to sort customers. Under the plausible assumption that those who are eager to watch a movie sooner are willing to pay more than those who are prepared to wait,[22] the revenue-maximising price structure displays a trajectory with prices falling over time. For instance, a movie is first released in theatres; with some lag it becomes available on a Pay-Per-View basis and/or DVDs. Next it may be released on a pay TV channel and ultimately shown on free TV. The staggered release of the movie through a sequence of windows achieves a price trajectory that has prices falling as waiting time to access the content increases. Consumer expectations matter in this context. If consumers thought that the content owners could not commit credibly to waiting six months after the theatrical release to releasing the DVD version, they might be less willing to pay the high price of the cinema version. The durability of content matters too. Blockbuster movies are generally regarded as more durable than sporting events, where the value to the consumer drops dramatically once the outcome is known (Cowie and Williams, 1997). Further, the segregation of the market into different windows is feasible only to the extent that leakage or piracy from one window does not undermine the revenue stream from subsequent windows. For instance, pirated copies created during the theatrical release would erode the revenue potential of the DVD and pay-per-view releases. Thus the optimal windowing strategy – for instance, the geographic extent and duration of release in each window and the price charged for that release – depends naturally on a variety of practical commercial considerations.

How does DRM improve the ability to charge discriminatory prices? Current windowing strategies allow only a handful of discrete windows, based on the delivery platforms (theatres, DVD, pay-per-view, and so on).[23] DRM can enable a finer partition of the access intervals – for instance, prices could be a continuous function of time elapsed since first release. By reducing the risk of leakage across windows, DRM can better segregate the different markets and can support higher prices in some windows.[24] Both of these are likely to enhance revenue.

Furthermore, DRM may allow the seller to discriminate access in dimensions other than just waiting time. In particular, where it is difficult to segment consumers according to exogenous characteristics, endogenous characteristics, such as the quality of the product being sold may be used to discriminate between consumers (Varian, 1997). By pricing the different quality of products appropriately consumers can be incentivised to self-select the price-quality bundle that reflects their willingness to pay. In this way, quality discrimination can be used to supplement price discrimination. A seller could discriminate between consumers who would like to watch a movie just once and those who are willing to pay for repeated viewings. A seller could restrict access by time of day or day of the week to distinguish between low-income students and busy executives. Similarly, a movie could be priced to discriminate between viewing technologies, with viewing in an

HDTV format used as a proxy for a higher willingness to pay than viewing in standard definition. In terms of the conventional typology of price discrimination, a significant proportion of the current approach can be viewed as analogous to third-degree price discrimination or group pricing: observably different groups of people are charged prices based on group characteristics like demand elasticity. DRM may replace this with a richer menu of carefully constructed consumption bundles, forcing individuals to select a bundle most suited to their privately known preferences.

At the same time, the ability of DRM technologies to monitor use may allow sellers to build better profiles of their consumers by tracking their past consumption patterns. While this may well conflict with privacy rules, it allows sellers to gain a better understanding of an individual's willingness to pay, and thus sort its customer more profitably for future transactions. Amazon.com bases its approach to making recommendations to users on its ability to track past consumption choices. It is not inconceivable that a similar approach could be taken to tracking consumption of audiovisual content within a pay TV environment. To the extent that prices can be conditioned on the history of consumption pattern, DRM may entail a step closer to the theoretical ideal for price discrimination, where every unit that a customer buys is charged at a price close to his willingness to pay.

What are the welfare implications of these changes? First-degree price discrimination generally enhances efficiency. Even though some consumers end up paying higher prices than they would under non-discriminatory pricing, the expansion of the market by enabling sales to previously excluded consumers is likely to be welfare-improving.

To understand this, imagine that consumers differ in more than one characteristic: their willingness to pay (one consumer is willing to pay more to see a James Bond movie than another) and their impatience (both would like to see the movie earlier rather than later, but one consumer is more impatient than the other). Windowing strategies exploit the typical correlation between these attributes, by selling at high prices to impatient consumers and low prices to patient ones. DRM could enable finer sorting of consumers even when these characteristics are not correlated. It could make the movie available to all consumers at an early stage, matching individual prices to their individual willing to pay, thereby increasing welfare. DRM succeeds here because supplying the movie cheaply to this person does not necessarily interfere with charging a higher price to other consumers. From the perspective of competition authorities concerned about the effects on the potential for competition of the exclusive contracts necessitated by inter-temporal price discrimination the alternative forms of price discrimination enabled by DRM may offer a more attractive market-driven outcome.

Even though there are strong theoretical arguments that suggest that better forms of price discrimination can potentially be welfare improving, it must be recognised that it generates considerable antipathy from consumers. In some cases, this even translates into calls for regulatory intervention. Groups such as the Electronic Frontier Foundation[25] have been particularly active in this regard, raising concerns about the potential for the breach of privacy rules, infringement of the consumers' alleged right of fair use and the potential for content companies to constrain consumers use of content unfairly (Electronic Frontier Foundation, 2004). What this activity illustrates is that not all parties in the supply chain agree that DRM provides the potential for an improved consumer experience. The challenge faced by the content industry is more complicated than competing with freely available pirate copies.

CONCLUSIONS AND IMPLICATIONS

Digital Rights Management may not be an entirely new concept, but the threat posed by piracy to the content industries, underpinned by the dramatic growth in broadband penetration, has brought debate on the technology and its implications to the fore. We have sought to look beyond the current debate on piracy to briefly predict how DRM may impact on business models in the content supply chain, in particular how DRM may potentially lead to a refinement of the traditional, rather clunky, windows-based content release model. What is clear is that DRM can cause an upheaval in the economics of the supply chain and a change in the way in which consumers access content, which may affect the way in which it is regulated. It remains to be seen how the incumbent platform operators will respond, but DRM does have the potential to achieve what competition authorities and regulators have thus far been unable to do, namely to reduce the market power of the vertically integrated pay TV platform operators and reduce the barriers to entry at key stages in the supply chain. The extent to which the pay TV bottleneck will simply be replaced by a DRM bottleneck remains to be seen, but the regulatory community will need to retain a close eye on this aspect of development.

NOTES

1. Shapiro and Varian (1998) describe information goods as 'anything that can be digitised'.
2. See Congressional Budget Office (2004).
3. See for example http://www.commsdesign.com/news/tech_beat/showArticle.jhtml? articleID=23901208.

4.   See the IT encyclopaedia web site, www.whatis.com.
5.   See Iannela, cited in Rump (2003).
6.   For more detail, see Becker *et al.* (2003).
7.   See http://www.commsdesign.com/news/tech_beat/showArticle.jhtml?articleID=23901208 for a recent discussion of watermarking technologies.
8.   Other works on the theme include Petrin (2002) and Nevo (2003).
9.   Deloitte (2004) estimate that the lost profits are in the range of $6bn to $7.5bn per year. Henning (2004) develops an econometric model to identify the variables that determine the likelihood of piracy for a given movie. The model estimates the cost of piracy to the German film industry at $153m in 2003. With extrapolation, this would suggest a cost of $3.6bn for the film industry worldwide. The gap between the two above estimates illustrates the difficulty in accurately estimating the cost of piracy to the film industry.
10. Article 6 of the Directive provides the framework for the protection of such measures against acts of circumvention.
11. The most significant decision in this regard is perhaps that which provided legal support for Sony's introduction of the Video Casette Recorder, which opined that the time-shifting enabled by the VCR amounted to fair use and therefore was not in breach of copyright law (*Sony Corp. v. Universal City Studios* 464 US 417 (1984)).
12. Additionally, the EU Copyright Directive reflects the view that parallel legislation on fair use and other exceptions will protect consumers. However, where voluntary measures by rights holders are deemed to be insufficient then the Directive empowers Member States to take remedial action, subject to the requirements of national legislation on fair use.
13. The determining variables in deciding whether to rely on intermediaries or to internalise the supply chain include transactions costs, scale and scope efficiencies, as well as asset-specific efficiencies.
14. See also Kaplan and Sawhney (2000), who explore the role of electronic intermediaries in the aggregation of demand and supply.
15. BSkyB reported profits of £154 million for the six months to December 2004.
16. www.movielink.com.
17. Issues such as standardisation, leverage of market power from related markets and patent pooling are beyond the scope of this chapter, but are none the less important areas for research.
18. As in Italy, Spain and France.
19. Indeed, in extreme cases, the standard model of monopsony (monopoly) may be more appropriate. If there is only one credible content licensee (licensor), the single pay TV channel (content supplier) would set the price for the content in a way that maximises its surplus.
20. The 'split-the-difference' solution of Nash Bargaining assigns to each player a payoff equal to his disagreement point plus a share of the surplus that remains after disagreement payoffs have been made. If outside options are viewed as disagreement points, they affect the Nash bargaining outcome directly. In contrast, the outside option principle of strategic bargaining argues that outside options merely constrain the set of possible solutions. If so, an outside option affects the outcome only if it enables a player to improve on the outcome in the absence of that option. For a discussion see Binmore *et al.* (1989), Muthoo (1999) and Osborne and Rubinstein (1990).
21. See for example, http://www.advanced-television.com/2002/June10_17.html. Peter Scudamore, CEO of the FAPL in 2002, said 'Next time round, rather than sell our rights to a broadcaster for them to sell on to households, the Premier League may do a deal direct with consumers, so that you can ring up [the] PL and say "I'd like to buy your Premier League channel with all the games on it for X pounds a month."'
22. The assumption that those who are most impatient are also those who place the highest value on the content is not always valid.
23. The gap that must be left between the windows to support the sequencing model has a cost in that the momentum of the movie's marketing is stalled at each stage. This may reduce the degree of self-generating publicity and require increased marketing spend.

24. To some extent the possibility of piracy is undermining windowing, as content released in one window is illegally transferred to consumers in ways that compromise the profitability of other windows. A rational reaction to this would be to increase prices in the early windows to compensate for compromised revenue in later windows, and possibly to compress the windows to reduce leakage of revenue streams (which we are seeing in regions where piracy is a major concern, such as Russia).
25. www.eff.org.

REFERENCES

Bailey, J. P. and J. Y. Bakos (1997) 'An exploratory study of the emerging role of electronic intermediaries', *International Journal of Electronic Commerce*, **1**(3), 7–20.

Becker, E., W. Buhse, D. Gunnewig and N. Rump (eds) (2003) *Digital Rights Management*, New York: Springer Verlag.

Besen, S. (1987) 'New technologies and intellectual property: an economic analysis', *Rand Report*, N-2601-NSF.

Bhargava, H. K. and V. Choudhary (2004) 'Economics of an information intermediary with aggregation benefits, *Information Systems Research*, **15**(1), 22–36.

Binmore, K., A. Shaked and J. Sutton (1989) 'An outside option experiment, *Quarterly Journal of Economics*, **104**(4), 753–770.

Brynjolfsson, E., Y. Hu and M. Smith (2003) 'Consumer surplus in the digital economy: estimating the value of increased product variety at online booksellers, *Management Science*, **49**(11), pp. 1580–1596.

Caves, R. (2001) *Creative Industries*, Cambridge, MA: Harvard University Press.

CBO (2004) 'Copyright issues in digital media', Congressional Budget Office Working Paper, August.

Committee on Intellectual Property Rights and the Emerging Information Infrastructure (2000) *The Digital Dilemma*, Washington, DC: National Academies Press.

Competition Commission (1999) *British Sky Broadcasting Group plc and Manchester United PLC: A Report on the Proposed Merger*, Cm 4305.

Cowie, C. and M. Williams (1997) 'The economics of sports rights', *Telecommunications Policy*, **21**(7), 619–34.

Deloitte (2004) *Pirates. Digital Theft in the Film Industry*, Deloitte.

EC Copyright Directive (2001) '*Directive 2001/29/EC of the European Parliament and of the Council of 22 May 2001 on the harmonisation of certain aspects of copyright and related rights in the information society*', *Official Journal* L 167, 22/06/2001, 10–19.

Electronic Frontier Foundation (2004) 'A better way forward: voluntary collective licensing of music file sharing, "Let the Music Play". White Paper'.

France Tribunal Paris (2004) 'Tribunal de grand instance de Paris 3ème chambre, 2ème section, Stéphane P., UFC Que Choisir/Société Films Alain Sarde et, Jugement du 30 avril 2004', available at: http://www.legalis.net.

Hausman, J. (1997) 'Valuing the effect of regulation on new services in telecommunications, *Brookings Papers: Microeconomics*, Washington, DC: Brookings Institute.

Henning, V. (2004) 'An empirical study of the effects of peer-to-peer filesharing on the film industry', Bauhaus University Weimar, Working Paper.

Kaplan, S. and M. Sawhney (2000) 'E-Hubs: the new B2B marketplace', *Harvard Business Review*, May–June, 97–103.

Lessig, L. (2004) *Free Culture*, Harmondsworth: The Penguin Press.

Liebowitz, S. (1985) 'Copying and indirect appropriability: photocopying of journals, *Journal of Political Economy*, **93**(5), 945–957.

Liebowitz, S. (2002), *Rethinking the Network Economy*, New York: Amacom.

Muthoo, A. (1999) *Bargaining Theory with Applications*, Cambridge: Cambridge University Press.

Nevo, A. (2003) 'New products, quality changes and welfare measures computed from estimated demand systems', *The Review of Economics and Statistics*, **85**(2), 266–275.

Novos, I. and Waldman, M. (1987) 'The emergence of copying technologies: what have we learned?', *Contemporary Economic Policy*, **5**(3), 34–43.

OFT (2002) *Decision of the Director General Fair Trading, BSkyB Investigation: Alleged Infringement of the Chapter II Prohibition*, CA98/20/2002, London: Office of Fair Trading.

Osborne, M. and A. Rubinstein (1990) *Bargaining and Markets*, San Diego, CA: Academic Press.

Petrin, A. K. (2002) 'Quantifying the benefits of new products: the case of the minivan', *Journal of Political Economy*, **110**, 705–729.

Rob, R. and J. Waldfogel (2004) 'Piracy on the high C's: music downloading, sales displacement, and social welfare in a sample of college students', *NBER Working Paper*, November.

Rump, N. (2003) 'Digital rights management: technological aspects', in E. Becker, W. Buhse, D. Gunnewig and N. Rump (eds) *Digital Rights Management*, New York: Springer, 3–15.

Samuelson, P. (2003) 'DRM {and, or, vs.} the Law', *Communications of the ACM*, **46**(4), 41.

Shapiro, C. and Varian, H. (1998) *Information Rules*, Cambridge, MA: Harvard Business School Press.

Sony Corp. v. Universal City Studios 464 US 417 (1984).

Tribunal Bruxelles (2004) Tribunal de première instance de Bruxelles, L'ASBL Association Belge des Consommateurs TestAchats/SE EMI Recorded Music Belgium, Sony Music Entertainment (Belgium), SA Universal Music, SA.

Tribunal de Grande Instance de Nanterre (2003) Tribunal de Grande Instance de Nanterre 6eme chamber Judgement du 2 September 2003, Francoise M./EMI France, Auchan.

Varian, H. (1997) 'Versioning information goods', University of California, Berkeley, mimeo.

Varian, H. (1998), 'Markets for information goods', University of California, Berkeley, mimeo.

10. Copy control of digital broadcasting content: an economic perspective

Koji Domon and Eulmoon Joo

INTRODUCTION

While music, photography, and home-videos have been digitalized, the remaining phase of media digitalization is broadcasting. Even though the Internet and media digitalization have developed concurrently, we cannot discuss broadcasting digitalization without accounting for the influence of the Internet. One such issue, the concern of content suppliers, is copyright infringement by Peer-to-Peer (P2P) systems. Broadcasters have seen music labels damaged by such infringements, and have managed to protect their content from it. To prevent content from outflowing into P2P networks, they have introduced copy control of digital broadcasting content, so-called 'Copy Once' in Japan and 'Broadcast Flag' in the US.

In this chapter, we shall consider, from an economic perspective, the copy control of digital broadcasting content and the resultant restrictions on personal use. In general, judgment of a copyright infringement is complicated. The concept of a copyright has a long history.[1] A basic mistake we sometimes make is treating copyrighted content as physical goods such as a pencil. Intellectual property[2] for the social welfare, including broadcasting content, has been treated differently from physical goods. Copying for personal use complicates this copyright issue. While photocopying and home video-recording make copying content free, content producers considered this copyright infringement at the time when such copy technologies had emerged. Legal judgment on this problem was that personal use of personally copied content was permissible with the US courts coining 'fair use' in dealing with the situation. The court decision was controversial since the exact effects of personal copying on the social welfare are difficult to estimate.

We shall consider this problem of personal use of digital broadcasting content from a transaction cost approach. Such an approach focuses on costs, except for a price of content, associated with a contract, transportation, a search for goods information and so on. Our analysis focuses on the transaction costs incurred from copying. This approach is not only familiar to economists, but is also understandable by those without economic knowledge. We proceed as follows. In the second section, we show how economics and 'law and economics' have analyzed, theoretically and empirically, a copyright of information goods. In the third section, we consider the problem of copy control of digital broadcasting content by using a numerical example and explain an 'indirect appropriability'. In the fourth section, we conclude with the effects on copyright holders and viewers of content.

AN ECONOMIC EXPLANATION OF COPYRIGHTED GOODS

Characteristics of Copyrighted Goods

Information goods, like broadcasting content, are called public goods in economics and are characterized twofold: non-rivalry of use among consumers and impossibility of excluding consumers from that use. Such goods are insufficiently supplied without legislative protection of usage, since content creators cannot obtain remuneration from others' pirate use of their content without it. Such a situation is socially undesirable.

Private goods, with both the rivalry and the possibility of exclusion, are protected by the law of property rights. Concerning this protection we need not worry about its use, since no one is allowed to thrive but can yet physically copy private goods. Furthermore, since the capitalism is based on property rights to private goods, we must protect these rights in order to sustain the market system. Compared to such rights, a copyright usage is ambiguous. A perfect exclusive right for creators is obviously not proper from a social perspective.

There are two factors complicating copyrights: incentive for creators and dynamic externality[3] that content provides for future creators. Regarding the first factor, the most profitable condition for a creator is a perpetual property right resulting in a perfect monopoly. However, such a monopolistic power incurs more social costs such as lesser consumer benefit resulting from higher prices and having a negative effect on future creation. Merely regarding incentives, it is difficult to determine the proper level of a copyright. Regarding the second factor, its effects are vague. All content, more or less, influences future creations. In other words, there is no creation which does not make use of a predecessor's outcomes. Such a phenomenon makes it extremely difficult to determine a proper copyright level.

A Fair Use and Development of Copy Technology

Historically, the first copyright, called a 'stationary right', was established for press owners in England. At that time press owners meant publishers. No one could produce a large quantity of books without a press. A copyright was infringed on by another press owner. However, nowadays, copyright infringement takes place among consumers due to the development of copy technology. As each new copy technology for easy and cheap copying emerged, we have had to reconsider the appropriate level of copyrighting. This permitted personal copying by photocopiers, cassette recorders, home-video recorders, and so on. In spite of differing restrictions between countries, the situation can be analyzed through a common economic factor, that is, a transaction cost.

In order to consider home video-recording of digital broadcasting content, an analysis of the *Betamax* case of *Sony vs. Universal City Studios* in the US courts provides a referential point. In this case, discussions were focused on 'fair use'[4] of copyrighted content. Gordon (1982) explained fair use in the *Betamax* case by using a transaction cost approach[5] that had been familiar to economists since Coase (1937). Exchange of goods, either at the marketplace or through contracts, incurs transaction costs other than those payments of goods. In ideal frictionless exchanges, that is, no transaction costs, a market force leads to a socially optimal situation. In reality, exchange of goods more or less incurs transaction costs which affect various economic behaviours.

In order for content copyright holders to charge a price for home video-recording, they must find and contract with those who record their content. Since such costs are usually prohibitively high, the transaction costs outweigh benefits from video-recording. These result in no market for a home video-recording service, since, when copyright holders can charge, no one buys content due to its higher price to that of benefit. In such a case copyright protection is not socially optimal, since benefits which, otherwise, could be obtained under free home video-recording are lost. In such a way the transaction cost approach explains the rationale for fair use.[6]

It should be noted that there are requisites for the above. First, the fair use must not significantly eliminate incentives for creators. Second, influence on market sales should be small. Third, it should contribute to social welfare. Supposing that the existing duration and extent of a copyright is proper, we should maintain it under the fair use. If these requisites hold, the fair use is socially beneficial even though personal copying prevails and copyright holders think that their rights have been infringed.

While this transaction cost approach to a fair use is persuasive for economists, it lacks concern for demand expansion. When Universal City

Studios fought Sony, it failed to account for the growth of the pre-recorded and rental video markets which were created by VCRs. After it lost the case, it noticed such markets and, as a result, both consumers and copyright holders benefited from VCRs. A similar situation may occur in copy-controlled media markets. Before analyzing this possibility, we will survey economics literature relating to copying.

Theoretical and Empirical Considerations

A paper by Ordover and Willig (1978) provides a theoretical analysis relating to the fair use of copyrighted goods. Although their consideration is photocopying of journals in a library, we can see the same critical factor, transaction costs, as seen by Gordon (1982) in the Betamax case. Users of journals decide whether to subscribe to them or photocopy them in the library, taking into account their transaction costs, that is, the inconvenience when they must use a library. Because users of a library[7] are restricted to the members of an institution, they indirectly pay a fee to copyright holders. In this case, the problem of copyright holders is that users share content in a library.[8] As a common feature of information goods, there is a large initial (sunk) cost for production compared to the marginal one causing a decreasing average cost. In a static situation, Ordover and Willig (1978) considered a benchmark for the Ramsey pricing for a social optimum.

Liebowitz (1985) considered copying in a library from a viewpoint which will be important to our analysis in the following section. He insisted that copying in a library is not harmful to copyright holders because benefits stemming from copying can be charged. He pointed out two kinds of appropriability of revenue: direct and indirect ones. A direct one takes place when a copyright holder charges purchasers for using content only for themselves. In this case, the charges depend on the purchaser's benefits. The second one occurs when purchasers permit others to copy their content. In this case a copyright holder can charge extra fees to purchasers due to the benefits from copying by others. As a result, he concluded that copying in a library was not a problem.

The two papers hinge on two assumptions in order for a copyright owner to do a price discrimination. The first is that suppliers obtain information on individual demand function. In a standard analysis of pricing, we assume that suppliers do not know the individual but rather the market demand function. Therefore, the first assumption is stricter than usual. The second is that consumers cannot do an arbitrage by using a price discrepancy between a high and a low price. Considering a transaction cost of the arbitrage, we may conclude that this assumption holds. The first assumption may also hold regarding journals supplied to a library, because suppliers appropriately know a library's demand from its scale.

In a static situation an incentive for copyright holders is profit from existing copyrighted goods. However, the incentive dynamically affects the future creation of content. Johnson (1985) focused on this issue and concluded that whether copying is harmful for society as a whole is unclear. He used the product differentiation model in which the producer's number, representing a variety of products, was variable. In the long-run, profit losses from copying cause an exit of producers from the market and a decrease in variety, while consumers' surpluses increase from copying.[9] Effects of copying on the social welfare depend on parameters of the model.

Besides these theoretical considerations, there are empirical ones which discuss the effects of P2P on market sales of music CDs. Liebowitz (2003) considered a historical trend of sales in the US, and concluded that the effects of P2P on revenues were negligible although sales of single CDs drastically decreased. Interestingly, he indicated that, with media developments, a demand curve for music content has shifted outward because the benefit of users has increased with them. That is apparent when we consider a media shift from vinyl records to CDs and finally the emergence of portable players. This phenomenon suggests that, in spite of illegal copying, an incentive to create content may continue since profits are barely affected by it. Oberholzer and Strumpf (2004) also support statistically the minor impacts of P2P on market sales.

We note that there are few theoretical or empirical papers which consider explicitly the dynamic externalities of copyrighted goods. A meaningful analysis relating to the accumulation of intellectual property or knowledge remains to be written.

A COPY PROBLEM OF DIGITAL BROADCASTING CONTENT

Characteristics of Digital Broadcasting Content

With digital technology we can easily copy various materials: a picture, an audio or a video. Digital broadcasting content is also copied on harddisk digital recorders. Before considering copy issues of the digital broadcasting content, we should consider its characteristics compared to other media.

Broadcast content can be viewed by those who have a TV set. However, viewers need to adjust their time to watch TV, but with a home video-recorder the value of content increases with the time-shift of watching. The value of content depends partly on how it is viewed.[10] A second characteristic relating to the above is geographical restriction of broadcasting. There are no channels, except on Internet TV, which are broadcast to everywhere in the world.[11]

These two characteristics are not at all related to content digitalization. How does digitalization affect viewers in a copyright context? The difference from the days of the Betamax case is that copying now can be done by file-sharing on the Internet. With Napster and Gnutella, the impact on the industry is much greater than with just a personal copy of an individual at home. In other words, if P2P were not popular, we would not have had to consider the matter of copying of digital content.[12]

Another phenomenon relating to content digitalization is Internet broadcasting in which some TV programs are now supplied by streaming. The impact of this copying of broadcast content will be significant, as it will lessen the impact of copying by P2P. Time-shifting will not be necessary, since content will be supplied on demand. Furthermore, we do not have to use P2P in order to obtain content broadcast in other areas, because we can access a web site from anyplace.[13]

Indirect Appropriability

In order to examine the effect of copy control on viewer and producer behaviours, we use an economic model to explain an indirect appropriability. How many times we copy is shown by the demand curve of copied content. For example, we sometimes photocopy (nowadays download) the same article several times to hold it both in the office and at home, or to store it for future reference. The best situation is to be able to use the content whenever and wherever we wish to at no cost. However, even though copying is free, we have to pay transaction costs for copying. Taking this into account, we determine in terms of proper copy times. Specifically, when the price of content is no more than the gross benefit, we purchase it.

Figure 10.1 shows a marginal benefit (a demand schedule) and the level of transaction costs for copying.[14] In this example, we suppose that a transaction cost to copy is $t = 30$. Facing such a situation, a producer has two options: free copy or no copy, as a normal CD or as a Copy Control CD (CCCD).[15] Responding to such, a consumer with the marginal benefit of Figure 10.1 has respectively the following benefits: Under free copy, his maximal payments (gross benefits) are the first marginal benefits plus benefits from copying:

$$100 + (80 - 30) + (60 - 30) + (40 - 30) = 190$$

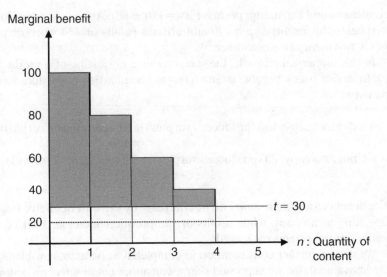

Figure 10.1   Optimal copy time

Table 10.1   Profit under no copy

Price	Profit
100	60
80	80
60	60
40	0

When a price under free copy is no more than those benefits, the consumer purchases content.

When free copying is prohibited, a consumer must purchase the amount needed. In such a case, the profit maximizing price is $p = 80$, when a production cost per unit is 40, which is in general larger than a transaction cost to copy. The profit is indicated in Table 10.1. Under no copy, a consumer purchases two products at $p = 80$ and obtains benefits:

$$(100 - 80) + (80 - 80) = 20$$

while the consumer obtains gross benefits of 190 under free copy.

If a price under free copy is less than 170 ($= 190 - 20$), the consumer obtains more benefits than when copying is prohibited. Therefore, corresponding to this consumer's behavior, the producer sets $p = 170$ under a free copy menu and can obtain profits, 130 (=170 − 40). As a result, a producer maximizes his profits at p =170 and obtains profits of 130 when he proposes two menus to a consumer.

In this numerical example, the social welfare, consisting of a producer's profit and a user's benefit, is also greater under a free copy than under no copy:

Under free copy: 130 (producer's surplus) + 20 (consumer's surplus)
= 150,
Under no copy: 80 (producer's surplus) + 20 (consumer's surplus)
= 100.

Net benefits for the consumer do not decrease by the shift of choice from a free copy to no copy, while profits of the producer increase. That is, the Pareto improvement is achieved.

We need a further consideration in completely solving this problem. In the above analysis, we supposed that a consumer could select an optimal menu from two choices. However, a producer does not have to offer the two. The highest profit is obtained by offering only a free copy. In this case the producer sets p = 190 and obtains profits, 150, because a consumer cannot select a free copy. The resultant social surplus is 150 (=150 (producer's surplus) − 0 (consumer's surplus)). Although social surpluses are at the same level as those under two choices, we do not obtain the Pareto improvement due to a decrease in the consumer's surplus. In any case, we are certain that free copy is the best strategy for a producer.

Effects of Copying and Copy Control

We will first restrict our consideration to copying of free TV content with commercials. When home video-recorders were unavailable, we had to adjust our time to a TV program schedule. In spite of free TV content, this adjustment charged us with implicit transaction costs. The effects can be explained using Figure 10.1.

We suppose that transaction costs to record TV content are at the same level as those to copy it. Under such a supposition, an increase to viewer's benefits from recording is in the gray area in Figure 10.1. Unless there is such an area, he is unwilling to record due to the higher transaction costs to benefits and he does not watch the TV program. Furthermore, we suppose that transaction costs to copy by P2P are less than those to record TV content.[16]

There is the possibility that personal copying by individuals at home has a higher commercial value for broadcasters due to an increase in viewing time. In such cases, as previously noted, personal copying is beneficial for both consumers and producers. A viewer obtains benefits represented by the gray areas in Figure 10.1, while the commercial value increases along with copy times.

However, this outcome becomes problematic with file-sharing[17] on the Internet, with no geographical restriction. It is apparent that consumers obtain more benefits from file-sharing because of lower copying transaction costs. Moreover, they can obtain TV content which they could not without file-sharing. We see that a geographical restriction gives a monopolistic power to a broadcaster. Competition beyond reachable areas of waves and signals decreases the profit level for each broadcaster. Incumbent terrestrial broadcasters have been threatened by newcomers such as CATV and satellite TV broadcasters. The same threat takes place under worldwide file-sharing of free TV content. This phenomenon cannot be considered in the analysis of an indirect appropriability. We cannot say generally which effect is greater, the negative effects from competition or the positive ones from the indirect appropriability.

Indirect appropriable revenues are also obtainable under personal copying of pay-TV content. In this case revenues are created by a higher price than that under copy control. This is possible because consumers' benefits increase by personal copying. A different point from free TV content is unpaid illegal copying. We consider this problem by comparison with unauthorized file-sharing of music tracks by P2P.

To compete with the file-sharing in music industries, major labels have shifted from sales in retail stores to online ones and have been successful as in the case of Apple's i-Tune. This success can be explained from a transaction cost approach. Consumers infringe on a copyright because of a no charge merit and a low transaction cost in spite of the risk of apprehension. Because consumers have differing assessments as to the risk of apprehension and transaction costs, some infringe while others do not. The success of online sales is attributed to lower transaction costs in order to obtain content which is more convenient on a PC and a mobile player than on a CD media player. These not only decrease infringers but also expand the market. An interesting strategy of online sales is the easy restriction on personal copying, which increases the merit for consumers when downloading is legal. Most consumers select the legal way, since unauthorized file-sharing burdens infringers with a relatively higher cost than with the benefit from infringement.

The unauthorized copying problem of pay-TV content can be analyzed in the same way as that of unauthorized file-sharing of music tracks. What corresponds to online sales of music tracks is video-on-demand service, supplied on CATV and webTV, which does not restrict the viewer to a time

Table 10.2   Effects of copy control on copyright owners

	Merit	Demerit
Free TV content	Protection of copyright Limited geographical competition among broadcasters	Losing a chance to expand the number of viewers
Pay TV content		Losing a chance to raise price by increasing benefits from copying

schedule. If a drastic price reduction of pay-TV content, as in music tracks, is possible, then producers do not have to fear unauthorized file-sharing. It is the inevitable side-effect of digitalized industries. One promising possibility of the price reduction is pay-TV on the web. Web TV producers need only servers, whose charges are much lower than the fixed ones of other media, to distribute content.

So far, we have considered the effects of file-sharing by P2P on copyright owners and viewers. Copyright owners can use copy control in order to nullify these effects. A summary of copy control effects on copyright owners is shown in Table 10.2.

CONCLUDING REMARKS

To prevent expansion of unauthorized copies of digital broadcasting content, various countries have introduced copy control schemes. However, these will not be effective, since illicit files are made on PCs with analog and/or digital TV tuners. These schemes do not cover PCs.

Furthermore, we must take into account the situation of the music industry. Major music labels in Japan announced that they would stop producing Copy Control CDs. According to them, CCCDs are now unnecessary since, by certain indictments,[18] they have eliminated unauthorized file-sharing by P2P. However, no one believes this. The truth is that CCCDs have lessened the benefits to users, as explained by indirect appropriability, and resulted in a decrease of CD sales. Another factor is the success of online stores and portable hard disk players. Sticking to CCCDs will not only bring a loss to profits, but also will exclude them from the market.

Compared to such a music industry, broadcasting now faces digitalization and the influence of the Internet which is now in the process of creating new distribution channels as well as new pricing for Pay-TV content.

In terms of unauthorized file-sharing, transaction costs for viewers to watch TV and illegal copying are the most important elements to be considered. In spite of any copyright law trying to cope with illegal copying, infringers will not soon disappear. The number of infringers is increasing, since they promptly use new technology as P2P while the response of producers takes time. The present situation of illegal copying reflects this. In such a situation, broadcasters must change the structure of the broadcasting industries as the music industries have done and are doing.

NOTES

1. The first concept corresponding to a copyright was a 'stationary right' in the UK. In the fifteenth century, stationeries pressing and selling books insisted on their exclusive rights to copy. Regarding the history of copyrighting, see Patterson (1968). Lessig (2003) also briefly refers to the history and offers an interesting insight on copyrighting in the Internet age.
2. Copyrighted goods are one of the intellectual properties covering a patent, a trademark, and so on. See Landes and Posner (2003) for an overall economic analysis of intellectual property rights.
3. In economics, externalities mean direct effects behaviors have on others outside a market system.
4. The fair use was enacted in Section 107 of the Copyright Act of 1976. See http://www.cetus.org/fair 5.html for a detailed explanation.
5. See Depoorter and Parisi (2002) regarding a brief discussion on the transaction cost approach.
6. Internet drastically reduces transaction costs and gives a producer a way to control content use. A content viewing, as well as software downloading, can be controlled, due to low transaction costs, on the Internet. In such a case, the fair use explained by a transaction cost will be meaningless, and the phenomenon as observed in music files purchased online can result. However, as long as existing broadcasting content is supplied mainly through terrestrial waves, controllability will be impossible and the fair use discussion will continue.
7. Although public libraries are free to use, the users indirectly pay a charge since they are funded by tax payers.
8. This type of shared goods is analyzed in a simple model by Varian (2000).
9. There are papers concerning the effects of a copyright protection on the social welfare. See Novos and Waldman (1984) and Conner and Rumelt (1991).
10. Even before a home video-recorder was invented, content value was influenced by functions of a TV set: a wide or a small screen, monaural or stereo audio, and so on. In such cases, consumers could control content usage. However, in Internet TV, suppliers can control the usage of content and discriminate between consumers. Online stores of music files have various menus by using controllability of the content usage. These phenomena mean that the controllability is shifting from users to suppliers.
11. When most subscribers of the Internet use a FTTH service, Internet TV will be as popular as Internet radio.
12. Although analog broadcasting content is shared under P2P by digitalizing it in an MPEG format, digital broadcasting makes file-sharing easier and P2P users can enjoy the merits of digital content. If viewers are satisfied with analog quality, copying by P2P is not problematic because the copying already exists and analog signals can be stored in PCs and distributed to P2P systems. Copy control of digital broadcasting content is implemented for only digital broadcasting content.

13. Illegal copying includes commercial piracy which will not be considered here. In Japan there are few cases of commercial piracy, while in the US it has caused significant damage. See Waterman (2004) regarding the US situation.
14. We can explain an indirect appropriability in a library by using this figure. If we can copy content freely in a library, we copy three times because the transaction costs are smaller than marginal benefits, and obtain the benefits of the gray area in this example. If producers know the benefits to users, they can add these to a normal price. In such a way, indirect appropriate revenues are generated by the gray area.
15. Music online stores, such as i-Tune, offer various options controlling copies: burning to CD, copy times to a hard-disk, streaming or storage, and so on. Taking these into account, we must consider many strategies for a producer. Such a consideration is beyond the scope of this chapter.
16. A condition for a viewer to record is that the first marginal benefit is greater than transaction costs to record. How can we explain behaviors of those who do not have an incentive to record but watch a TV program? Their transaction costs to adjust their time to watch TV is less than their transaction costs to record. In other words, their opportunity costs of time are relatively low. Moreover, those who do not watch TV programs on time and by a home video-recorder have a relatively high opportunity cost of time and transaction costs to record.
17. In P2P networks, we can find broadcasting content which deletes commercials. If most content in P2P networks were such, there would be no indirect appropriable revenues from copying. However, this situation will not take place because deleting content incurs costs. Most content will be supplied with no cuts.
18. An effect of indictments is the increase of transaction costs, a risk of apprehension, for infringers, which decreases the number of infringers.

REFERENCES

Coase, R. H. (1937) 'The Nature of the Firm', *Economica*, **4**, 386–405.
Conner, K. E. and R. P. Rumelt (1991) 'Software piracy: an analysis of protection strategies', *Management Science*, **37**(2), 125–139.
Depoorter, B. and F. Parisi (2002) 'Fair use and copyright protection: a price theory explanation', *International Review of Law and Economics*, **21**, 453–473.
Domon, K. and N. Yamazaki (2004) 'Unauthorized File-Sharing and the Pricing of Digital Content', *Economics Letters*, **85**(2), 179–184.
Gordon, W. J. (1982) 'Fair use as market failure: a structural and economic analysis of the Betamax case and its predecessors', *Columbia Law Review*, **82**, 1600–1657.
Johnson, W. R. (1985) 'The Economics of Copying', *Journal of Political Economy*, **93**(1), 158–174.
Landes, W. N. and R. A. Posner (2003) *The Economic Structure of Intellectual Property Law*, Cambridge, MA: Harvard University Press.
Lessig, L. (2003) *Free Culture*, New York Times.
Liebowitz, S. J. (1985) 'Copying and indirect appropriability: photocopying of journal', *Journal of Political Economy*, **93**(5), 945–957.
Liebowitz, S. J. (2003) 'Will MP3 downloads annihilate the record industry?: the evidence so far', in Gary Libecap (ed.), *In Advances in the Study of Entrepreneurship, Innovation, and Economic Growth*, New York: JAI Press.
Novos, I. E. and M. Waldman (1984) 'The effects of increased copyright protection: an analytic approach', *Journal of Political Economy*, **92**(2), 236–246.
Oberholzer, F. and K. Strumpf (2004) 'The effect of file sharing on record sales: An Empirical Analysis', mimeo.

Ordover, J. A. and R. Willig (1978) 'On the optimal provision of journals qua sometimes shared goods', *American Economic Review*, **68**(3), 324–338.

Patterson, L. R. (1968) *Copyright in Historical Perspective*, Nashville, TN: Vanderbilt University Press.

Takeyama, L. N. (1994) 'The welfare implications of unauthorized reproduction of intellectual property in the presence of demand network externalities', *Journal of Industrial Economics*, **62**(2), 155–166.

Varian, H. R. (2000) 'Buying, sharing and renting information goods', *Journal of Industrial Economics*, **48**(4), 473–488.

Waterman, D. (2004) 'The political economy of audio-visual copyright enforcement', mimeo.

PART III

Digital broadcasting and platform competition

11. Regulation of digital TV in the EU: divine coherence or human inconsistency?

Luca Di Mauro

INTRODUCTION

The EU regulatory framework for digital TV is based on two main pillars: regulation of content and regulation of infrastructure. These two pillars have different origins, aims at different objectives, and are implemented through different legislative means.

Regulation of content in the field of television broadcasting has primarily been based on the Television Without Frontiers Directive (TVWF). Despite its title, and while its purpose is allegedly to create a single market in the production and distribution of audiovisual content, the Directive introduces a number of restrictions on the provision of audiovisual content. In particular, the Directive has been conceived to support objectives which do not seem to be particularly relevant to, or instrumental to, the objective of developing a single market for broadcasting content: among them, for example, ensuring cultural and linguistic diversity quotas, the protection of minors, of human dignity and of consumers.[1]

Regulation of the infrastructure needed to convey broadcasting signals has been, on the other hand, increasingly shaped by the phenomenon of convergence. Convergence of technologies, allowed by the adoption of digital signal standards for data transmission, has meant that regulation of infrastructure capable of conveying digital data has had to evolve and adapt. Regulation of this kind has primarily been developed in the EU within the context of telecommunications. Today such regulation has been enshrined in the EU regulatory framework for electronic communications.

The distinction between hardware and software therefore remains one of the fundamental characteristics of European regulatory intervention in the audiovisual field. This chapter, however, will strictly focus on the former type of regulatory intervention. Even without accepting the view, which the writer retains, that almost all types of content regulation are largely unnecessary and mostly ineffective,[2] it is difficult to deny that economic regulation can and should only affect the structure of the market if it is targeted at the infrastructure-level and not at the content-level.[3]

Hence the use of the word 'regulation' in this chapter is referred solely to economic regulation of the infrastructure needed to convey digital TV broadcasting signals. In this respect, I will first provide an overview of the regulatory framework applicable to digital television (DTV). I will then attempt to identify the determinants of such framework, and in particular the economic background to it. I will finally provide an assessment of whether the current framework, as derived from its relevant theoretical background, is fit for purpose, with a discussion of the type of objectives on which regulatory intervention in this complex field should focus, before drawing some conclusions.

REGULATION OF TECHNICAL SERVICES FOR DIGITAL TV IN THE EU

DTV has been only recently introduced throughout the EU. It was first launched in 1996, first on satellite and soon after on cable and terrestrial networks, based on Digital Video Broadcasting (DVB) specifications. The main driver behind digital TV take-up has been so far satellite pay-TV, with free-to-air still accounting for less than 20 per cent of total digital TV viewing, although rapidly rising. In turn, pay-TV has been driven by multi-channel and premium programming.

In short, DTV allows for a much more efficient use of the main resource used to deliver TV signals, that is, spectrum frequencies. The greater efficiency of digital technology applies across the range of media which can be used to convey the signals: cables of different material or simply the atmosphere. Hence the desire of legislators to support the switch to the generalised use of digital technology, which in turn poses a number of questions as to what the most appropriate measures are to support such a switch. One of the fundamental questions is how to regulate the infrastructure used to deliver broadcast signals: different types of regulation can lead to radically different outcomes and can have a substantial impact on the speed with which the switch can be achieved.

In the EU, technical and economic regulation of digital TV has been primarily addressed by way of the introduction of Directive 95/47/EC,[4] which has been written with three objectives in mind: first, creating the legislative basis for a harmonised introduction of new TV services; second, encouraging the use of a number of standards and mandating standard-enforcing mechanisms; and third, laying the grounds for regulation of the rapidly evolving infrastructure used to deliver TV services.

Of these objectives, the latter is surely the most important one for the purposes of this chapter. As far as Directive 95/47/EC is concerned, regulation of the DTV infrastructure mainly means regulation of conditional access systems. Conditional access systems (CAS) are defined in Article 2(f) of the Framework Directive as 'any technical measure and/or arrangement whereby access to a protected radio or television broadcasting service in intelligible form is made conditional upon subscription or other form of prior individual authorisation'. This means that a CAS allows encryption, or otherwise scrambling, of the broadcast signals so that only those who are entitled to receive them can do so. In doing so, a CAS represents perhaps the single most important mechanism to distribute a TV signal over a network of viewers, which hence can become users of a platform.

The introduction of CASs means a radical departure from what has been the more traditional broadcasting mode since the inception of television, be it via terrestrial or via cable: the shift from a public good (or quasi-public good) status to that of standard good. Public goods are products or services characterised by two conditions – first, that persons cannot be excluded from consuming a product or service, and second, that one person's use of a product or service does not diminish another person's ability to use it. These two characteristics can be separately defined as non-excludability and non-rivalry, respectively.

It is clear that TV broadcasts, in the way every citizen has known them in the last half of the last century, have been non-excludable and, to a large extent, non-rival. Once TV signals are broadcast over the airwaves, any individual who possesses the appropriate infrastructure can access them without any further level of control. Even on a cable infrastructure it is theoretically possible to intercept the signal at any given point without the possibility for the provider of excluding individual viewers. A CAS allows TV broadcasters to select their viewers to a degree which was not known before.

CASs therefore represent one of the pillars of DTV services. By allowing for a much greater degree of control over what kind of information individual viewers can access, they provide broadcasters and platform operators with the capability to target their products much more selectively and experimenting with their material more innovatively. Regulation of conditional access is therefore of crucial importance in respect of the development of TV services. Article 4 of Directive 95/47/EC introduces the basic rules applying to all providers of conditional access:

Article 4

In relation to conditional access to digital television services broadcast to viewers in the Community, irrespective of the means of transmission, the following conditions shall apply:

(a)   all consumer equipment, for sale or rent or otherwise made available in the Community, capable of descrambling digital television signals, shall possess the capability:

  (i)   to allow the descrambling of such signals according to the common European scrambling algorithm as administered by a recognised European standardisation body,

  (ii) to display signals that have been transmitted in clear provided that, in the event that such equipment is rented, the rentee is in compliance with the relevant rental agreement;

(b)   conditional access systems operated on the market in the Community shall have the necessary technical capability for cost-effective transcontrol at cable head-ends allowing the possibility for full control by cable television operators at local or regional level of the services using such conditional access systems;

(c)   Member States shall take all the necessary measures to ensure that the operators of conditional access services, irrespective of the means of transmission, who produce and market access services to digital television services:

  (i)   offer to all broadcasters, on a fair, reasonable and non-discriminatory basis, technical services enabling the broadcasters' digitally-transmitted services to be received by viewers authorised by means of decoders administered by the service operators, and comply with Community competition law, in particular if a dominant position appears,

  (ii) keep separate financial accounts regarding their activity as conditional access providers.

Broadcasters shall publish a list of tariffs for the viewer which takes into account whether associated equipment is supplied or not. A digital television service may take advantage of these provisions only if the services offered comply with the European legislation in force;

(d)   When granting licences to manufacturers of consumer equipment, holders of industrial property rights to conditional access products and systems shall ensure that this is done on fair, reasonable and non-discriminatory terms. Taking into account technical and commercial factors, holders of rights shall not subject the granting of licences to conditions prohibiting, deterring or discouraging the inclusion in the same product of:

  (i)   a common interface allowing connection with several other access systems, or

  (ii) means specific to another access system, provided that the licensee complies with the relevant and reasonable conditions ensuring, as far as he is concerned, the security of transactions of conditional access system operators.

Where television sets contain an integrated digital decoder such sets must

allow for the option of fitting at least one standardized socket permitting connection of conditional access and other elements of a digital television system
of the digital decoder;

(e) without prejudice to any action that the Commission or any Member State
     may take pursuant to the Treaty, Member States shall ensure that any party
     having an unresolved dispute concerning the application of the provisions
     established in this Article shall have easy, and in principle inexpensive, access
     to appropriate dispute resolution procedures with the objective of resolving
     such disputes in a fair, timely and transparent manner.

(f) This procedure shall not preclude action for damages from either side. If the
     Commission is asked to give its opinion on the application of the Treaty, it
     shall do so at the earliest opportunity.

What is the gist of the provisions contained in this article? It introduces a
minimum level of regulation in the market for the provision of conditional
access services. This minimum level of regulation is relatively well-known in
the economics of competition and of regulation: it entails obliging a
provider to offer its services on terms which are fair, reasonable and non-
discriminatory (FRND).

The origins of this level of regulation must be looked for in the all-
embracing debate on convergence which has populated European policy
arenas in the 1990s. At that time, the clear question for policy-makers was
how to respond to the advent of digital technology and to the impact it would
have had on the IT, media and telecommunications industries. While it was
not at all clear that a response was needed, the debate eventually polarised
around two positions: at one end were the supporters of the creation of two
different frameworks, one for broadcasting and the other for telecommunications, while on the other hand it was argued in favour of a horizontal
framework which would cover both industries.[5]

Broadcasting remained characterised, much more than telecommunications, by disparate national rules and procedures covering matters such
as licensing, media ownership and control, foreign ownership, and media
content. Telecommunications, on the other hand, had just started bene-
fiting from the process of liberalisation, where national monopolies were
slowly giving way to growing competition, with clear effects for end users,
at least in terms of price decreases.

The European Commission eventually wrote its Green Paper on
Convergence,[6] whose stated objective was to address in an open and
objective manner the appropriate direction of regulation for the converging communications environment. The key question posed by the Green
Paper was how to achieve regulatory consistency in an environment where
services can retain their specific characteristics regardless of their means
of conveyance. Although the Green Paper did not take a clear position
on which approach would be best suited to regulating the converging

industries, and although it did not go so far as to actually recommend the creation of a single regulator for those industries, it can be argued that it did contain a bias towards horizontal (that is, cross-sectoral) regulation.

Most commentators argued that it was necessary to build on existing regulation rather than opting for a revolutionary approach.[7] In their view, broadcasting would have continued to have a unique relevance for the expression and formation of opinion, and therefore a 'fundamental role . . . for ensuring democratic pluralism, diversity and the sharing of culture'.[8] For obvious reasons, the same respondents did not see the telecommunications industry as a key player in the democratic pluralist arena, which should have led to its continued distinctive sector-specific regulation.

With this background, and after a number of other policy initiatives and working papers, the European Commission opted for a mixed approach. In fact, in its new regulatory framework for electronic communications the European Commission, Parliament and Council have set out the principles for regulation of all converging infrastructure – including traditional telephony, access to the Internet and broadcasting transmission services. However regulation of content, as previously mentioned, is exempted from the application of a common framework and, apart from the norms contained in the TVWF Directive, is still largely dependent on national policies.

The new regulatory framework (NRF) entered into force on 25 July 2003 and consists of four Directives[9] and a number of accompanying measures[10] which replaced the many legislative measures on which regulatory intervention in the sector had been based in the past. It brings about significant changes both as regards the scope and role of regulation of the electronic communications industry in Europe and as regards the role played by competition policy. The new framework applies *ex ante* regulatory intervention using concepts and principles taken directly from standard competition law theory and practice, thus aligning existing sector-specific regulation on general principles of competition law.

Under the old framework, national regulatory authorities (NRAs) could impose regulatory obligations on undertakings which had a market share of 25 per cent or more. Furthermore, *ex ante* regulation was focused on certain markets which were not 'markets' within the meaning of competition law but rather 'market areas' based on specific policy considerations.

The NRF is instead based on antitrust principles and, therefore, *inter alia*, technologically neutral. The trigger for regulatory intervention is a level of market power which, unlike the static criterion of 25 per cent market share used previously, must be assessed based on a complex economic analysis centred upon the competition-law based notion of dominance.

NRAs carry out an in-depth market analysis for all the markets listed in the Commission Recommendation on relevant markets[11] and must then conduct both a 'national' and a 'Community' consultation on the measures they intend to take. They can then inform their decision on whether maintaining, amending or withdrawing regulation depending on the results of the market analysis. The approach chosen in the NRF heavily draws on the successful and innovative approach to regulation adopted by the UK regulator, Oftel, since the beginning of the 1990s, which put economic analysis at the very heart of regulatory policy and thus created a much-needed link between the previously mainly administrative approach to regulation and antitrust enforcement.

One of the key features of the NRF is that NRAs can assess market conditions in principle only for those markets which have been defined as relevant for the purposes of *ex ante* regulation by the European Commission.[12] The markets in the Recommendation are identified as those in need of *ex ante* regulation, in addition to standard antitrust enforcement. Hence regulatory intervention becomes a two-step process, whereby first relevant markets are identified, and second, the level of market power of companies active in those markets is assessed. It is clear, then, that the initial choice of markets susceptible to *ex ante* regulation is of paramount importance. This choice, according to the European Commission, is made on the basis of three criteria which would indicate whether a market can be left to standard and horizontal antitrust law, or if instead its structure is such that a higher level of regulatory intervention is needed.[13] Effectively, this implies that the scope of regulatory intervention on electronic communications markets across European Member States is to a large extent defined through the choice of markets included in the Recommendation.

How does the NRF touch upon broadcasting infrastructure? The NRF does not formally distinguish between different types of electronic communications infrastructure, hence broadcasting infrastructure is in principle treated in the same way as any other telecommunications infrastructure. However in substance some key differences emerge. In particular, the NRF moves in two different directions when it comes to broadcasting. First, it identifies the market for broadcasting transmission services as one of the relevant markets for the purposes of regulatory intervention:[14] hence NRAs must carry out an analysis of the structure of the market and assess the level of market power of the companies operating in it. Second, it incorporates into the framework the same provisions on technical services applied to digital broadcasting which were first introduced with the *Advanced Television Standards Directive*: namely, conditional access and other ancillary services, including access control and access to Electronic Programme Guides. In particular, Article 5 of the Access Directive establishes that:

1. National regulatory authorities shall, acting in pursuit of the objectives set out in Article 8 of Directive 2002/21/EC (Framework Directive), encourage and where appropriate ensure, in accordance with the provisions of this Directive, adequate access and interconnection, and interoperability of services, exercising their responsibility in a way that promotes efficiency, sustainable competition, and gives the maximum benefit to end-users.

In particular, without prejudice to measures that may be taken regarding undertakings with significant market power in accordance with Article 8, national regulatory authorities shall be able to impose:

(a) to the extent that is necessary to ensure end-to-end connectivity, obligations on undertakings that control access to end-users, including in justified cases the obligation to interconnect their networks where this is not already the case;
(b) to the extent that is necessary to ensure accessibility for end-users to digital radio and television broadcasting services specified by the Member State, obligations on operators to provide access to the other facilities referred to in Annex I, Part II on fair, reasonable and non-discriminatory terms.

The 'other facilities' referred to above are access to application programme interfaces and access to Electronic Programme Guides. Article 6 of the same Directive, on the other hand, is entirely dedicated to conditional access services. In referring to Annex I to the Directive for the specific rules to be applied to conditional access providers, Article 6 introduces again exactly the same provisions seen above, which stem from the *Advanced Television Standards Directive*. The rest of the Article attempts to reconcile the rules laid down in respect of all other digital infrastructure in the NRF on the one hand, with the clear exceptions represented by digital broadcasting on the other:

3. Notwithstanding the provisions of paragraph 1, Member States may permit their national regulatory authority, as soon as possible after the entry into force of this Directive and periodically thereafter, to review the conditions applied in accordance with this Article, by undertaking a market analysis in accordance with the first paragraph of Article 16 of Directive 2002/21/EC (Framework Directive) to determine whether to maintain, amend or withdraw the conditions applied.

Where, as a result of this market analysis, a national regulatory authority finds that one or more operators do not have significant market power on the relevant market, it may amend or withdraw the conditions with respect to those operators, in accordance with the procedures referred to in Articles 6 and 7 of Directive 2002/21/EC (Framework Directive), only to the extent that:

(a) accessibility for end-users to radio and television broadcasts and broadcasting channels and services specified in accordance with Article 31 of Directive

2002/22/EC (Universal Service Directive) would not be adversely affected by such amendment or withdrawal, and
(b) the prospects for effective competition in the markets for:
   (i) retail digital television and radio broadcasting services, and
   (ii) conditional access systems and other associated facilities, would not be adversely affected by such amendment or withdrawal.

The next section will provide an interpretation of these provisions and will attempt at defining the theoretical background to them.

TRACING BACK THE DETERMINANTS OF DIGITAL BROADCASTING REGULATION IN THE EU

The FRND provisions stem from the experience in the application of competition law, as they mainly entail using the principle of non-discrimination, which is seen as generally characterising a behaviour which is not anticompetitive: under established antitrust case law, certain price discrimination strategies, when put in place by dominant undertakings, are deemed to be abusive. However, and more in general, these provisions belong to the class of prices known as access prices.

The problem of setting appropriate access charges to an existing infra-structure (which can be a network, a pipe, or, in theory, a broadcasting platform) has been extensively studied in economic literature. In general, such literature focuses on the situation where a large, complex, very expensive and not easily replicable facility has been deployed by a monopolist company. A substantial body of work exists both in relation to vertically unbundled industries, and in relation to industries where the incumbent owner of the facility is vertically integrated, and could therefore have the incentive to distort prices and raise competitors' cost.[15]

Where the owner of a facility which cannot be easily replicated also competes downstream with a user of the bottleneck service, the obvious concern is that it may have the incentive to foreclose entry to the facility or to set access charges which are high enough to make entry difficult or to limit competition. However, from a public policy perspective, the opposite case where the access price is too low is also undesirable, as under those circumstances inefficient entry may occur. In general, entry is inefficient when its positive effect on consumer surplus minus its (negative) effect on industry profits is not sufficient to justify the entry costs.[16]

In all those industries where the costs of the infrastructure, or the facility, are particularly high, economic analysis has shown that the 'first best' of pricing access at 'marginal cost' is infeasible. Traditional marginal cost pricing would under those circumstances lead to lower output and potentially to exit from the market. In these cases access prices should include a contribution to the fixed cost of the infrastructure, since new entrants should contribute to repay the fixed cost of the service that they use. The costs to which this contribution is sought are known as 'common costs', as opposed to costs which are specific, or incremental, to individual services/products. There can be different ways to recover common costs, which can take into account, for example, the price elasticities of the access seekers (Ramsey pricing, which leads to greater mark-ups over marginal costs in inelastic segments of the market), or the direct cost of providing access plus the opportunity cost of providing access (Efficient Component Pricing Rule or ECPR).[17] All these approaches recognise that under the circumstances discussed here new entrants need to contribute towards the fixed cost of infrastructure, and that access charges should be set at a higher level than the direct short-run marginal cost (or the avoidable cost) of providing access. In this context pricing above marginal cost is not an exercise of market power, but merely the efficient manner of financing socially beneficial investments.

It is clear that this is the theoretical background which should have been taken into consideration when looking into the issue of conditional access charges. It is less clear that this was the case: the FRND rule does not provide a clear access pricing method nor can it always be interpreted uniquely. This is the reason why a number of countries have explored ways to deal with the difficulties raised by the FRND provisions, and in particular with the fact that although a certain degree of discrimination has been found to be not only necessary but also beneficial to competition, those rules may be read as disallowing discrimination *tout court*. As mentioned above, economic analysis clearly recognises that price discrimination can be welfare-enhancing in many contexts,[18] and in particular in industries where companies incur very high levels of common costs to build an infrastructure needed to deliver a number of different products.[19] This is clearly the case with broadcasting markets, as well as with most electronic communications markets.

From an economic perspective it seems sensible to interpret the FRND requirements in line with the standard objectives of competition and regulatory policy. Competition and regulatory policy must aim at ensuring that the competitive process is fostered as much as conceivably possible across all markets within an industry. This outcome is widely recognised as the best guarantee that overall welfare is maximised. Within this context, non-discrimination rules are normally seen as a helpful tool only under specific circumstances.

From this perspective, the need for a non-discrimination obligation arises where the conditional access provider is also vertically integrated and provides its input to a broadcasting company. A vertically integrated provider may have an incentive to provide products on terms which discriminate in favour of its own operations. In the best of all worlds, therefore, a non-discrimination obligation must be aimed solely at ensuring that vertically integrated providers of conditional access do not treat themselves on preferential terms in such a way as to have a distortionary effect on competition. The practical objective at which to aim should therefore be to ensure that the conditional access provider offers products in such a way that the alternative broadcasters seeking access to the platform run by the conditional access provider are placed in an equivalent position to the platform operator's own channels.

The UK regulator Oftel (now Ofcom) offered perhaps the most complex and advanced interpretation of the FRND terms along these same lines: Oftel would have considered 'whether the terms offered are consistent with those which would be expected in a competitive market'.[20] In a competitive market a company would expect, among other things, to be offered conditional access at prices which are somehow consistent with the underlying costs ('fair'), timely and with the necessary informational and practical support ('reasonable'), and at rates which are equivalent to those offered to similar companies ('non-discriminatory').

Oftel has published a series of consultation documents and guidelines to explain how it interprets the requirements to supply on (FRND) terms, and in particular (a) what costs should be recovered by the service provider through their total level of charges, and (b) how these costs should be allocated between different types of users. The Oftel/Ofcom guidelines have consistently upheld three main principles: first, that in the presence of economies of scale and scope, access charges should include a contribution to common costs;[21] second, that common costs should be borne by access seekers on the basis of the differential willingness-to-pay of different customers (access seekers); and third, that outcomes that are negotiated commercially are preferable and that the regulator should only get involved following a complaint.

In particular, even when a provider of conditional access adopts a strategy of subsidisation of set-top boxes, or incurs other investments in its digital platforms, Oftel found that, unless cost recovery by the conditional access provider was specifically found to have anti-competitive effects, 'other broadcasters typically benefit from such a subsidy (of consumer equipment) (in terms of increased viewer base), so it is therefore reasonable to expect them to contribute to the costs.'[22]

In making a charge for this cost, Oftel also made clear that it would take account of uncertainty, and would not punish companies for a short-term over-recovery that could not have reasonably been predicted at the time prices were set. This uncertainty was partly taken into account by allowing that revenues and costs need not necessarily balance out in any single year, but rather that the net present value of prices versus costs should not be excessive over the entire period of the project.[23]

Also, the UK regulator has made it very clear that it does not interpret the 'non-discrimination' requirement set out in the FRND provisions to imply that all users must be charged exactly the same price. Instead, Oftel established that any differences in prices must be 'objectively justifiable'.[24] Oftel has acknowledged that the most efficient way to set prices (in order to distribute common costs between end users) should take into account the end users' willingness to pay for the different channels. Hence channels with higher revenues would contribute more to fixed and common costs than channels with more limited revenues (which would pay closer to their incremental cost). This principle has been accepted since the earliest days of digital conditional access regulation in the UK. Hence, and very reasonably, Oftel/Ofcom has not proposed a specific access pricing formula for linking access charges to retail prices, but has stated explicitly that an economically efficient pricing structure should reflect the willingness to pay in the price charged for conditional access services, as indicated, for example, in the retail price, or subscription price, or advertising revenues of a channel (or package of channels).[25]

This approach reflects the consideration that broadcasters of premium channels may have a greater willingness to pay for the services and be less likely to exit the market in response to a higher charge (as they receive a greater retail price for their channels, and therefore have more to lose if they are barred from access). At the same time, such a pricing structure enables the provider of access services to recover a greater share of fixed and common costs from established premium channel providers than from start-up or free-to-air channels (who might be more likely to exit in response to a higher charge), which ultimately and arguably thereby encourages greater variety of channels and greater choice for viewers.

The FRND provisions must then be read in the context of access pricing, and it appears sensible to interpret them in light of the above mentioned objective of supporting the development of a competitive market. However it may not always be possible to interpret them in such a benign way. More specifically, two questions arise: first, can conditional access charges, and charges for technical services, be non-discriminatory? Second, should those charges be non-discriminatory? The next two sections will try to answer these questions in turn.

DISCRIMINATING WITHIN A NON-DISCRIMINATION ENVIRONMENT

A strictly legalistic interpretation of the FRND provisions would be likely to completely distort the economic concepts which underlie beneficial regulatory intervention in the broadcasting markets. Strictly speaking, the FRND terms impose on a company the obligation not to discriminate. Depending on who the judge is who is called to assess the 'fairness' and 'reasonableness' of charges, it would be possible to get a different consideration of the specific terms of access offered by the conditional access provider. When assessing the non-discrimination requirement, however, it may well be possible for an impartial judge to come to the conclusion that no discrimination of any kind should be allowed, and that charges by conditional access service providers must be exactly the same for all users of those services.

This conclusion, as seen above, would run counter to the established wisdom based on the same economic analysis which should underpin regulatory policy. Regulation aims, or should aim, at establishing a framework to provide incentives for an efficient use of limited resources and to promote, wherever possible, a competitive market. The FRND terms should therefore be interpreted in a way that ensures that the market for conditional access services, and thus, more generally, the market for TV services, can prosper by coupling economic efficiency with innovation, dynamism and high quality of products and services. While any interpretation of FRND may need to be reviewed in the light of the changing characteristics of the market and the needs of the players involved, any interpretation should also create the conditions for the market to develop.

The fundamental problem is that it is impossible not to discriminate in setting prices for the provision of conditional access and other technical services. At the same time, choosing a different pricing method may have a substantial effect on the overall costs borne by (both access and incumbent) operators, and thus, ultimately, on the state of competition in the TV markets.

In fact, there are no clear reasons to prefer one pricing solution to another in terms of economic analysis. This is because the provision of conditional access and technical services generates a level of common costs which is extremely high relative to that of incremental costs. The common costs are recovered by mark-ups over incremental costs. These mark-ups can be calculated in a number of ways, for example in terms of an amount per channel, or per package, or as a proportion of retail revenue, or on some other basis. Whichever method is chosen is likely to disadvantage some types of operator against others. Hence the reason for the impossibility not to discriminate is that when a high level of common costs needs to be recovered, any chosen allocation criteria would result in some degree of discrimination.

For example, single per-channel charges would result in broadcasters offering packages of many channels paying much more than broadcasters offering few channels towards the recovery of fixed common costs. This would happen despite the underlying incremental costs not being greatly different. Such a pricing structure would discriminate in favour of those broadcasters which offer one channel targeted at many customers, as such broadcasters would pay less than broadcasters offering many channels accessed by few customers. The negative side of this approach would be that it would provide a relative incentive to single-channel providers to access the platform, rather than to multi-channel providers. This could ultimately lead to under-utilisation of existing platform capacity and to a less-than-optimal degree of output, which would in turn adversely affect end users' welfare.

Or, charges could be set per broadcaster, so that the incentives provided would be exactly the opposite: the access cost per channel would be lower for broadcasters with many rather than with few channels, even though, as already noted, the incremental costs of the services offered by both types of broadcaster would not, in fact, greatly differ. This approach would be justified if it were found that the willingness to pay of an N-channel broadcaster is not N times the willingness to pay of a single channel broadcaster: this would be the case, for example, if each channel at the margin were less lucrative. This would also support some kind of non-linear charging schedule, though not necessarily a fixed per-package charge. However such an approach would be likely to deter entry from single-channel providers, as well as encouraging channel providers to merge and consolidate, potentially leading to a situation where new channels/content is taken up by existing providers rather than offered by new entrants. This may ultimately lead to a greater degree of concentration in channel provision, which may well be at a level beyond what would otherwise be expected or indeed desirable.

Alternatively, charges could be set based on the retail price of the channel, or package of channels, of the access seeker. The rationale for this approach is that it would directly take into account the consumer's (and, as derived from this, the broadcaster's) willingness to pay, as measured by the retail price of the package. The same approach could be used by taking into account the acceding channel's advertising revenues. This pricing method would encourage entry of providers of free-to-air services as well as less expensive retail packages, thereby potentially increasing the number and variety of services available. It would also eliminate the problem of the access fee acting as a potential barrier to entry to the platform, since smaller channels with low revenues would not be priced out of the market. However, and clearly, a charge which varies with the retail price of the access seeker will discriminate between them, as broadcasters which charge different prices will pay different access prices.

Hence any type of conditional access charge seems to be discriminatory to some extent. The same applies to all other technical services currently regulated under the NRF, as access to APIs and to EPGs. Hence, in order to make it possible for the FRND provisions to be interpreted in a way which does not contradict established economic findings, as well as common sense, it is necessary to adopt a wise and flexible approach to the application of the non-discrimination requirement.

This is indeed what has been done in the UK, and what may become more commonplace across the EU if the same approach prevails. Oftel/Ofcom has adopted a flexible approach which is loosely based on Ramsey-type principles and whereby it is recognised that economically efficient prices would reflect neither the number of channels nor the number of operators, but would have to reflect mainly the relative willingness to pay of customers. In particular, Oftel/Ofcom interprets the FRND requirements according to the following principles: first, on average the conditional access operator should be able to recover its costs and make a return on its investment which is appropriate to the level of risk and uncertainty at the time of investment. Second, prices for particular categories of services should fall between the incremental cost of providing that service and the total cost of providing that service on its own. And third and foremost, comparable users are charged comparable prices for comparable services at equivalent points in time, with a view to ensuring that vertically integrated suppliers do not supply to their own downstream operations on more favourable terms than those offered to third parties.

This approach is flexible enough not to penalise operators which incur a high level of investment costs in order to set up a platform. Also, this approach takes the view that conditional access providers should enjoy a certain degree of freedom to negotiate different pricing arrangements with third party service providers in order to maximise usage of the system, whilst also ensuring there are no material adverse effects on competition. This is thought to have promoted an economically efficient utilisation of the platform.

The answer to the first question asked at the end of the last section, therefore, must be that a strict interpretation of the FRND-type of regulation would not seem to allow for any degree of discrimination, but that such an interpretation would run against established economic principles and results, would be likely to cause under-investment in new infrastructure and under-utilisation of existing resources, and it would ultimately lead to an inefficient outcome in the broader TV markets.

However a second question arises, when stepping back and looking with a fresh pair of eyes at the EU regulatory framework on digital TV infrastructure: is regulatory intervention really needed and, if yes, is the current level of regulation appropriate?

WHY REGULATE?

Modern economic analysis suggests that regulatory intervention is needed only in those circumstances where market failures occur, or where the likelihood of companies exploiting their excessive level of market power is high. These principles have been embodied in the NRF. Since the 1999 European Commission Communication[26] which preceded the launch of negotiations over the four Directives of the NRF, a few key principles have been identified as characterising regulatory intervention. In particular, regulation should be based on clearly defined policy objectives; be the minimum necessary to meet those objectives; further enhance legal certainty in a dynamic market; aim to be technologically neutral; and be enforced as closely as possible to the activities being regulated.

Hence modern regulatory policy envisages devising a pro-competitive environment in the markets which are subject to *ex ante* regulatory intervention. While in the past regulation was at times considered as a synonym for a fragmented and inconsequential set of norms, which might have eventually led to a situation where the development of competition was held back rather than supported, in more recent times regulatory action has been increasingly based on the same set of tools on which competition analysis is based. Regulation is hence essentially economic regulation, and economic regulation is based on the perspective that intervention on the market is necessary and beneficial only when it offers the solution to certain sorts of market power. In particular, regulation is considered necessary to remedy specific market failures, which often derive from formerly monopolistic market structures. However, regulatory action is not meant to be a self-perpetuating instrument, but rather an instrument to allow, as much as conceivably possible, competition to develop.

The NRF therefore introduces a dramatic change with respect to the previous regulatory framework. While a very limited number of NRAs have substantial experience in imposing remedies which are targeted at specific competition problems, the vast majority of them were used to the rather more 'automatic' mechanics set out in the previous framework. According to the Open Network Provisions (ONP), the basis for the previous regulatory framework, most regulatory intervention was meant to ensure inter-connection of electronic communications infrastructure at regulated prices, without a detailed explanation of how to construct the link between assessment of the competitive conditions in a market and imposition of regulatory obligations. In other words, one could have been forgiven for interpreting regulatory intervention as a purely administrative matter within the ONP framework.

The NRF, on the other hand, provides a comprehensive toolbox to address competition problems in a very flexible way, with proportional responses and based on economic analysis, as opposed to administrative mechanics. One of the latest additions to the NRF, the ERG Common Position on appropriate remedies,[27] clearly shows the fundamental shift to proportional responses to competition problems and a pre-eminent role of economic analysis in regulation. Under the NRF, NRAs must carry out market analyses and choose one or more appropriate remedies from a non-exhaustive list of regulatory obligations, starting from the lowest level (obligation of transparency) to the highest level (price control obligations). Regulation is now considered as a complementary tool to antitrust enforcement, and each competition problem is dealt with by focusing on its source rather than on its effects. This in general coincides with the objective of targeting wholesale markets when trying to remedy competition problems which may appear evident only at the retail level.

Why regulating electronic communications markets *ex ante*, then, rather than leaving them to antitrust enforcement? The reason is that standard antitrust tools rely on a set of assumptions which are not satisfied in electronic communications markets. In particular, they rely on the assumption that markets are characterised by the simultaneous presence of a large number of firms, which did not previously enjoy strong market positions and have grown more or less organically: the reference point is the model of perfect competition.

This is clearly not the case with the electronic communications industry. In particular it is not the case with the most traditional telecommunications markets, which have been characterised in all Member States by the presence of monopolies which have been charged with the exclusive remit of providing telephony services to whole countries. It is of course entirely possible to face the consequences of the unregulated behaviour of these incumbents in each Member State through standard competition law. However there are at least four reasons why it would not make much sense to deal with such problems under standard competition law. First, the sheer number, intensity and frequency of the abuses which would be likely to occur would make it impossible for a competition authority to be effective in dealing with them. Second, competition authorities would find it very difficult to respond quickly enough to the competition problems arising, hence intervention would not prevent the exacerbation of those problems. Third, competition authorities would lack the necessary technical knowledge of the sector, which is essential in order to understand the nature of the problems. Fourth, standard competition law would not be able to impose detailed appropriate and detailed obligations, since the remedies available to it are not as flexible and scalable as those normally available under *ex ante* regulation.

Regulation in general, and the NRF more specifically, cannot therefore be seen outside of the specific context from which it originated. This context is the process of liberalisation, which in itself carries the marks of a difficult question: how to open markets to competition, which were originally conceived as markets to be served through publicly owned and publicly run companies. The aim of regulation is twofold: creating a pro-competitive environment on the one hand, while seeking to provide, on the other hand, the benefits to end users which the market would provide if it were effectively competitive. The aim of creating a pro-competitive environment is best served through a regulatory framework which, among others, ensures the creation of sustainable competition over time. This is ultimately a function of the type, quality and quantity of infrastructures available. The aim, on the other hand, of providing end users with the benefits of an effectively competitive market is best served, among others, with ensuring the highest possible degree of competition in each potentially competitive segment of the industry, by, for example, opening up previously monopolistic markets to competition through the imposition of access obligations.

In particular, the objective of creating a pro-competitive environment can be seen as centred around the objective of providing incentives for competition based on alternative facilities. On the other hand, the objective of recreating a market outcome similar to that which end users would get from an effectively competitive market can be seen as centred around the objective of providing incentives for competition based on the service provision model. The latter is a model of competition based on an incumbent providing access to its own infrastructure, and competitors offering services at the retail level based on the access products of the incumbent.

The objective of providing incentives for alternative operators to build their own infrastructure, which is consistent with the first set of objectives above, would appear at first sight to be incompatible with the objective of ensuring that end users benefit as much as possible of a market outcome which mimics the outcome of an effectively competitive market. This is because rolling out networks can take a very long time and involves high levels of cost. If regulation aims at recreating effectively competitive market conditions in an artificial way, that is, by imposing access obligations and without substantially changing the underlying market structure, incentives would not exist for alternative operators to build their own infrastructure.

Regulation of digital TV infrastructure in the EU clearly must be seen in this context. The FRND terms represent a specific level of regulation in the list of possible regulatory interventions which are available to NRAs.[28] While the obligation of access is formally included in the Framework Directive as a separate obligation,[29] imposing the obligation to provide services at FRND terms on a vertically integrated provider *de facto* means imposing an access obligation on that provider.

This happens in the context of the NRF, which has been reportedly conceived to provide a comprehensive, consistent and harmonised set of rules to apply to all digital communications infrastructure, regardless of the technology used and regardless of the markets served. In a context where NRAs are required to assess the level of market power of undertakings in each market for which they are competent, including the market for broadcasting transmission services, and impose on dominant companies a remedy which must be proportional to its level of market power, the NRF then goes on to establish that for conditional access services, as well as for technical TV services, NRAs can impose an FRND-type obligation on any provider of those services, regardless of its degree of market power as well as of any preliminary assessment of how needed such an intervention is. The question arises of whether such regulation of the market for technical TV services is justified and, if yes, whether it is appropriate.

Looking at the structural characteristics of these markets, it is difficult to understand the justification for this approach. Digital TV services are regulated asymmetrically with respect to other electronic communications services. However, and crucially, they are regulated asymmetrically only insofar as they are always regulated irrespective of the degree of market power of companies providing those services, and not in respect of the type of regulation. The type of regulation imposed is entirely based on a telecommunications paradigm, where providers are seen as vertically integrated and some form of access obligation is imposed to 'open up the platform'.

However most electronic communications markets present certain characteristics, mainly related to the role of former public monopolies in them. On the opposite, the market for TV services, as well as the TV industry in general, seems to be quite different from the traditional telecommunications markets in many respects.

First, the operation of a digital TV infrastructure is essentially a fixed cost business. While this is similar to telecommunications, unlike the latter, digital TV markets are not characterised by the presence of huge national businesses which have been entrusted with the remit to provide universal telephony services to the entire population, which has been the model throughout the EU in the past century. Digital TV platforms, on the opposite, have been mainly characterised by organic growth (with some exceptions, in particular for some of those countries where cable operations have been preferred to terrestrial operations), and in any case have greatly benefited from pay-TV services as a key driver for their growth. Everywhere in the world, pay-TV services have been the realm of private ventures wishing to offer targeted services to their audiences, rather than the realm of former public monopolies which have been subsequently privatised. Pay-TV operations have often created new markets which did not exist before, rather than operating in monopolistic markets subsequently liberalised and opened up to competition. The difference with respect to incumbent telecommunications operators could not be greater.

Second, the more traditional telecommunications markets are more clearly characterised by the presence of a less competitive segment of the industry, normally identified with the 'last mile' and with parts of the backhaul network, together with the presence of a prospectively competitive segment of the industry, normally identified with services to specific users (for example business users) or to specific locations (for example large metropolitan cities). On the opposite, digital TV markets can be characterised by a much greater degree of vertical integration. Broadcasters remain broadcasters regardless of the technological platform on which they operate (at least so far). However digital TV platforms which are also broadcasters may choose to be vertically integrated because a tighter control over the value chain is seen as beneficial to the success of the platform. The infrastructure is rolled out in order to allow end users to have access to content: hence in the long term most of the market power in the markets, if any, comes from content (rights) holders, rather than from the infrastructure owner. Even in those cases where a pay-TV undertaking has deployed a very large number of set-top boxes, two conditions would seem to be a prerequisite for regulatory action: first, it should be demonstrated that a similar deployment could not be matched by other undertakings; second, it should be demonstrated that by mandating access on a platform, the objective of achieving a more sustainable level of competition would not be more difficult to achieve.

Third, and consequently, the model of competition which applies to digital TV services appears to be very different from that which applies to telecommunications. While a large part of electronic communications markets are characterised by a structure which can support, and benefit from, high levels of competition without the overall industry suffering from it, there is some evidence to show that in digital TV markets the structure may be such that more competition downstream, both on the platform or between platforms, always implies greater competition for access to content.

This, in turn, may lead to situations of financial struggle for some operators and may ultimately lead to one or more operators not being viable. This is likely to be more true when two or more operators compete using the same type of technological platform,[30] thus being more constrained in relation to rights acquisition: technological neutrality is unfortunately an analytical principle which still has not much following in the business world. While such scenarios may not arise, competition between platforms for the same content may result in end users not being able to be provided a full range of choice and being forced to renounce to some content or to face high switching costs to have access to alternative platforms.

Fourth, even if competition between digital platforms were deemed to be a desirable process, the FRND-type of regulation introduces a distortionary element in the competitive dynamics, as it allows for a degree of intra-platform competition. However if the objective is to achieve sustainable competition, the actual concern should be inter-platform competition. Sustainable competition in electronic communications markets is normally seen as related to platform competition.[31] Content access obligations may well have no effect on the roll-out of alternative platforms, or may slow down that process.

Hence there is a twofold inconsistency in the regulation of digital TV services in the EU. First, some of these services (that is conditional access and other technical services) are always regulated, without any apparent consideration of the criteria set out to assess whether these markets present problems which warrant regulatory intervention, while other digital TV markets (that is broadcasting transmission services) are subject to the standard rules of the NRF. Second, the type of regulation envisaged for conditional access services seems to be inappropriate or anyway typical of another context, that of electronic communications services. No suitable analysis has been apparently carried out to support the view that the current type of regulation of those services is beneficial to society.

If the real concern were that of remedying a market failure or responding to a high degree of market power, it would seem more appropriate to treat digital TV services as any other electronic communications service, since the NRF allows for a proportionate response to each competition problem. Therefore if there were competition problems in digital TV markets, they could be addressed similarly to any other competition problem under the NRF: by looking at the source of the problem and devising a proportionate response. Alternatively, it would seem more appropriate to encourage a public debate on the model of competition which is envisaged for digital TV markets, with a view to devising, if any, the more beneficial type of regulatory intervention for these markets, including standard *ex post* antitrust enforcement.

However all seems to point to the fact that the actual concern behind the imposition of regulation based on FRND principles does not lie in a market power or competition problem but elsewhere.

CONCLUSIONS

The actual concern behind the choice of adopting an FRND remedy for digital TV services described above may well be that digital broadcasting is seen as similar to all other electronic communications services. However the fact that such regulation is applied regardless of the degree of market power of operators points to the fact that digital broadcasting services are also considered unique and apparently deserve special treatment. Since the FRND remedy can be implemented in very different ways, the risk is that more stringent interpretations of the remedy are adopted in different countries as a function not of a solid economic analysis, but of more volatile factors, perhaps related to the political or broadcasting context in Member States. Such ambiguity undermines the robustness of the EU regulatory framework when applied to digital TV.

Broadcasting has a stronger tradition of policy intervention than other information and communication sectors, where the impact of liberalisation has been greater. This is justified, in the view of many observers and policy-makers, by the political and social relevance of broadcasting content, which calls for the enforcement of minimum quality and pluralism requirements. Policy intervention has been greater in the case of analogue terrestrial broadcasting because of its heavy use of spectrum, a scarce public resource, as well as the perception associating terrestrial with universal free-to-air TV services.

However, the contexts surrounding the introduction of analogue and digital broadcasting are very different. When analogue broadcasting was introduced, only the terrestrial option existed. There was no competition and the market was entirely shaped by regulatory intervention. In terms of economics, it can be argued that the very first deployment of TV services implied high and absolute barriers to entry and a very high level of sunk costs, which may be an argument in favour of publicly funding the roll-out of analogue TV services. Today, on the other hand, there are various types of networks, using different types of technology, which in turn has led to a higher degree of competition and to even faster technological change. If anything, a debate on which model of competition, hence of regulation, is the most appropriate one for digital TV services should take place. Rather than that, the EU has been discussing the wonders of convergence for almost 20 years, has been devising a harmonised and internally consistent regulatory framework for another five years, only to end up treating digital TV, as usual, as the only exception which warrants asymmetric regulation.

The theoretical background to this differential treatment seems to be a different concept of market structure than that applied in other electronic communications markets. When it comes to digital TV, the main concern seems to be that end users are made captive by all-powerful media platforms, who control the 'bottleneck' to them and hence can prevent other broadcasters from having access to them.

However this approach is slightly inadequate since it makes use of categories of analysis (such as the catch-all word 'bottleneck', which almost always is used to call for regulatory intervention on any market entity depending on the context). The real problem is, as it always is, the level of market power possessed by firms in any market. A firm's degree of market power may well be greater because of the fact that such a firm owns a facility which cannot be easily replicated, or a 'bottleneck' facility. However this will only be one of the factors in assessing its degree of market power, to be considered together with all other relevant factors, and taking account of any specific circumstances and of the relevant model of competition envisaged for the market.

The focus of policy-makers and regulators, therefore, should not be on access *to* end users,[32] which should be granted to broadcasters, but on access by end users to as many broadcasters as possible. Viewers cannot be seen as a resource, or an essential facility, controlled by a digital TV platform. They must instead be seen as market agents who need to be empowered and put in a position to make choices which will determine the broader structure of the market.

The risks inherent in the current approach to regulation of digital TV services are that unjustified regulation is not only inconsistent with the objectives of the NRF, but can also be detrimental to the development of these markets. Public intervention in the provision of digital TV services can be misused and abused, for example by industrial parties seeking to offset commercial risk, thus leading to reduced competition and diminished pressure to innovate. This in turn can result in perverse effects, like moral hazard or market inaction, and ultimately slow down the growth of these markets and of publicly stated objectives such as the switchover process. Parties may be encouraged to mix private and collective benefits and get support by public authorities in the name of the general interest to gain a competitive edge over rivals.

It is clear that the perception exists that the broadcasting sector is not comparable to any other sector, as it plays a central role in modern democratic societies, notably in the development and transmission of social values.

Broadcasting offers a unique combination of features. Its widespread penetration provides almost complete coverage of the population across different broadcasting networks and it allows for the provision of substantial quantities of news and current affairs which, together with cultural programming, lead to both influence on and reflection of public opinion and socio-cultural values.

At the same time, digital TV has been hailed as an era of change, of greater choice for end users and of greater variety of offerings. It would be quite unfortunate if the EU could not take advantage of what is possibly the most advanced regulatory framework in the world for this specific sector, hence preventing regulation of digital TV services from pursuing the same objective as regulation of other sectors of the economy: putting the end user at the centre. A reasonable interpretation of FRND regulation appears to be absolutely vital to help this objective come true.

ACKNOWLEDGEMENTS

I am grateful to Martin Cave for very helpful comments on the chapter and Tommaso Valletti for fruitful discussions on two-sided markets. I would also like to thank Peter Culham and Geoffrey Myers for helping me to develop my thinking on regulatory policy. However, and obviously, the views expressed in this article, as well as any mistake, are only mine and they do not necessarily reflect the views of my Organisation.

NOTES

1. Such objectives are considered by the European Commission as being of paramount importance for the general interest: see European Commission (2003).
2. In particular, content regulation appears to be at best anachronistic, and at worst distortionary, when it is applied to digital content. Such content is traditionally characterised by the possibility to use forms of non-linear consumption.
3. Of course content regulation can have deep effects on the structure of the market under certain circumstances: for example, when an undertaking retains monopoly power of content, regulatory intervention can substantially alter market structure.
4. Directive 95/47/EC of the European Parliament and of the Council of 24 October 1995 on the use of standards for the transmission of television signals, *Official Journal* L 281, pp. 51–54.
5. See Sauter (1998).
6. European Commission (1997).
7. European Commission (1998).
8. European Commission (1998), p. 4.
9. Framework Directive – Directive 2002/21/EC of the European Parliament and of the Council of 7 March 2002 on a common regulatory framework for electronic communications networks and services, OJ L108, 24.4.2002, p. 33. Access Directive – Directive

2002/19/EC of the European Parliament and of the Council of 7 March 2002 on access to and interconnection of electronic communications networks and associated facilities, OJ L108, 24.4.2002, p. 7. Universal Service Directive – Directive 2002/22/EC of the European Parliament and of the Council of 7 March 2002 on universal service and users' rights relating to electronic communications networks and services, OJ L108, 24.4.2002, p. 51. Authorisation Directive – Directive 2002/20 of the European Parliament and of the Council of 7 March 2002 on the authorisation of electronic communications networks and services, OJ L108, 24.4.2002, p. 20.

10. Amongst them the Radio Spectrum Decision – Decision 676/2002 on a regulatory framework for radio spectrum policy in the European Community, OJ L108, 24.4.2002, p. 1.

11. Commission Recommendation 2003/311/EC of 11 February 2003 on relevant product and services markets within the electronic communications sector susceptible for *ex ante* regulation in accordance with Directive 2002/21/EC of the European Parliament and of the Council of 7 March 2002 on a common regulatory framework for electronic communications networks and services, OJ L114, 8.5.2003, p. 45.

12. NRAs can deviate from the list of markets identified in the Recommendation on relevant markets, by either analysing markets not included in the list or assessing that a market included in the list does not warrant *ex ante* regulatory intervention. However, any such deviation needs to be justified in the light of the three criteria which the Commission used to identify the markets to be subject to *ex ante* regulation (see above, note 3).

13. The first criterion is whether a market is subject to high and non-transitory entry barriers, the second criterion is whether a market has characteristics such that it will tend over time towards effective competition and the third criterion is whether competition law is sufficient on its own to address the perceived market failures absent *ex ante* regulation.

14. Market no. 18 in the Recommendation on relevant markets is defined as 'Broadcasting transmission services, to deliver broadcast content to end users'.

15 For a detailed survey, see Armstrong (2002).

16 See, for example, Mankiw, N.G and M.D. Whinston (1986).

17 For a more detailed discussion see for example, Baumol and Sidak (1994); Economides and White (1995); or Laffont and Tirole (1996).

18. See Varian (1989) for a broad discussion of price discrimination and of its welfare properties.

19. See Varian (1985) for a discussion of price discrimination strategies for multiproduct firms with high levels of common costs.

20. Oftel (2002), p. 7

21. Oftel (1999), A.62.

22. Oftel (2002), 3.10.

23. Oftel (1997), A.74.

24. Oftel (2002), 2.9.

25. Oftel (2002), 3.7.

26. European Commission (1999a). The same principles are restated in European Commission (1999b).

27. European Regulators Group (2004).

28. In particular, the non-discrimination obligation is included in the Framework Directive as Article 10 and represents the second step in a ladder of five categories of obligations, from transparency (lowest) to price control (highest). While the obligation of terms being 'fair' and 'reasonable' are not explicitly included in the list of obligations in the Framework Directive, they can be seen as *de facto* included, among others, in the transparency obligation.

29. Article 12 of the Framework Directive.

30. This has been the case, for example, in Italy, where the two platforms which preceded to Sky Italia, Stream and Telepiù, had been incurring losses for 5 and 8 years respectively before the European Commission allowed their merger. The combined platform is expected to break even in 2006.

31. See for example European Regulators Group (2004), p. 13.

32. As is reflected in Annex I, Part I of the Framework Directive, point (b).

REFERENCES

Armstrong, M. (2002), 'The theory of access pricing and interconnection', in M. Cave, S. Majumdar and I. Vogelsang (eds), *Handbook of Telecommunications Economics*, Amsterdam: North-Holland.

Armstrong, M., S. Cowan and J. Vickers (1994) *Regulatory reform – Economic Analysis and British Experience*, Cambridge (MA): MIT Press.

Baumol, W. J., and J. G. Sidak (1994) 'The pricing of inputs sold to competitors', *Yale Journal on Regulation*, **11**(1), 171–202.

Canoy, M., P. de Bijl and R. Kemp (2002) 'Access to telecommunications networks', Paper prepared for the European Commission, DG Competition.

Cave, M., S. Majumdar, H. Rood, T. Valetti and I. Vogelsang (2001) 'The relationship between access pricing and infrastructure competition', Report to OPTA and DG Telecommunications and Post, Brunel University.

Cave, M. (2002) 'Remedies in network industries: competition law and sector-specific legislation. An economic analysis of remedies in network industries', mimeo, Brussels, 26 September.

Cave, M. (2003) 'Remedies for broadband services', Paper prepared for the European Commission.

Council of the European Union (1989) Directive 89/552/EEC on the co-ordination of certain provisions laid down by law, regulation or administrative action in Member States concerning the pursuit of television broadcasting activities, 3 October.

Economides, N. and L. J. White (1995) 'Access and interconnection pricing: how efficient is the efficient components pricing rule?', *The Antitrust Bulletin*, **XL**(3), 557–79.

European Commission (1997) *Green Paper on the Regulatory Implications of the Convergence of the Telecommunications, Audiovisual and Information Technology Sectors*, Brussels, December.

European Commission (1998) *Working Document of the Commission: Summary of the Results of the Public Consultation on the Green Paper on the Convergence of the Telecommunications, Media and Information Technology Sectors: Areas for Further Reflection*, Brussels, July.

European Commission (1999a) *Towards a new framework for Electronic Communications infrastructure and associated services – The 1999 Communications Review*, Communication from the Commission to the European Parliament, the Council, the Economic and Social Committee, and the Committee of the Regions, COM (1999) 539.

European Commission (1999b) *Principles and guidelines for the Community's audiovisual policy in the digital age*, Communication from the Commission to the European Parliament, the Council, the Economic and Social Committee, and the Committee of the Regions, COM (1999) 657.

European Commission (2003) *Communication from the Commission to the Council, the European Parliament, the European Economic and Social Committee and the Committee of the Regions on the Future of European Regulatory Audiovisual Policy*, Brussels, COM (2003) 784, 15 December.

European Parliament and Council of the European Union (1995) 'Directive 95/47/EC of the European Parliament and of the Council of 24 October 1995 on the use of standards for the transmission of television signals', *Official Journal* L 281, 0051–0054.

European Regulators Group (2004) *ERG Common Position on the Approach to Appropriate Remedies in the New Regulatory Framework*, ERG (03) 30rev1.

Laffont, J. and J. Tirole (1996) 'Creating competition through interconnection: theory and practice', *Journal of Regulatory Economics*, **10**(3), 227–56.

Laffont J. J., P. Rey and J. Tirole (1998) 'Network competition I: overview and nondiscriminatory pricing', *RAND Journal of Economics*, **29**, 1–37.

Mankiw, N. G. and M. D. Whinston (1986) 'Free entry and social inefficiency', *RAND Journal Of Economics*, **17**(1), 48–58.

Office of Fair Trading (1999b) *Competition Act 1998 – The Application in the Telecommunications Sector*, London.

Office of Telecommunications (1999) *The Pricing of Conditional Access and Access Control Services: Ensuring Access on Fair, Reasonable and Non-discriminatory Terms*, London, May.

Office of Telecommunications (2002) *Terms of Supply of Conditional Access: Oftel Guidelines*, London, March.

Perry, M. K. (1989) 'Vertical integration: determinants and effects', in R. Schmalensee and R. Willig (eds) *Handbook of Industrial Organization*, vol.1, Amsterdam: North Holland.

Sauter, W. (1998) 'EU Regulation for the Convergence of Media, Telecommunication, and Information Technology: Arguments for a Constitutional Approach', ZERP-Diskussionspapier 1/98, Zentrum für Europäische Rechtspolitik an der Universität Bremen, pp. 12–26, p. 21.

Varian, H. R. (1985) 'Price discrimination and social welfare', *American Economic Review*, **75**, 870–875.

Varian, H. R. (1989) 'Price discrimination', in R. Schmalensee and R. D. Willig (eds), *Handbook of Industrial Organization*, vol. I, 597–654.

Valletti, T. (2003) 'Obligations that can be imposed on operators with significant market power under the new regulatory framework for electronic communications', paper prepared for the European Commission.

12. Platforms for the development of digital television broadcasting and the Internet

Hajime Oniki

INTRODUCTION

In Japan, digital television (DTV) for terrestrial broadcasts was introduced in December 2003. Due to historical, political, and other reasons, the introduction of DTV is considered merely as a replacement for analogue with digital content; there has been little discussion regarding its impact on business practices and industry structure. Accordingly, the benefits of DTV are said to be those of technical improvements such as spectrum saving, noise prevention, finer images (HDTV), and multi-channel capability.

The impact of the digitization of television, however, will reach far beyond those technical improvements for at least two reasons. First, it can increase viewer satisfaction by expanding their choices with regard to the timing of watching programmes.[1] Further, it is now possible to increase the usefulness of the content to consumers by processing it with computer and storage technologies. DTV programmes may be used and reused, with possible modifications, for educational, cultural, business, and other activities.[2] The potential benefits from this are so great that it is impossible to imagine them at present.[3]

Second, DTV has provided the possibility for television to compete and/or co-ordinate with the Internet. In short, the Internet is a system for transmitting digital information on a global scale. DTV is a system for broadcasting digital information. It is evident that DTV and the Internet can and should work closely together for the benefit of society. However, because of certain historical reasons, DTV and the Internet at the present time are still two mutually exclusive systems; competition and co-ordination between DTV and the Internet is yet to occur.

This chapter deals with the impact of the introduction of DTV in Japan and in other countries, along with the possibilities of using the power of computer and storage technologies, and with DTV's competition and co-ordination with the Internet. It will focus on the need for, and the implication of, forming 'platforms' for digital business, a way to establish 'efficient division of labour' in the digital world, in relation to DTV. The chapter will also attempt to identify political, legislative, and regulatory impediments to the smooth formation of such platforms and to suggest policy recommendations for overcoming them.

DIGITAL BROADCASTING AND THE INTERNET IN JAPAN

DTV for terrestrial broadcasting in Japan was started in December 2003. By 2011, terrestrial broadcasting for traditional analogue television will be terminated according to a schedule set by the Japanese Ministry of Internal Affairs and Communication (MIC). It is uncertain, however, whether a majority of television viewers in Japan will or will not purchase DTV receivers by that time, or whether broadcasters and cable-TV providers will or will not be able to deploy DTV networks to cover most areas in Japan by that deadline. MIC has not made clear whether the use of spectrum blocks for analogue television broadcasting would be terminated in the case that the purchase of receivers for, and the deployment of, DTV falls behind the scheduled time limit. MIC has made plans to subsidize television broadcasters operating in rural areas with regard to deploying DTV services; however, no plan has been made to assist consumers, or regulate manufacturers, so as to promote the purchase of DTV receivers by consumers (even in the case in which consumers replace worn-out analogue receivers).

Although DTV started earlier than 2003 with other media such as communications satellites (in 1996, called CS broadcasts), cable TV (in 1998), and broadcasting satellites (in 2000, called BS broadcasts), its introduction into terrestrial television broadcasting is considered to have a major impact, since its revenue in 2002 represented nearly 80 per cent of all revenue for the broadcasting industry (including cable).[4] The per capita (per consumer) annual revenue of the industry is approximately JPY30 000 (US$250), with the average length of time devoted by Japanese households to watching TV exceeding 3 hours per day.[5] Furthermore, terrestrial television is probably the most influential medium on the Japanese public in political, social, and cultural arenas.

The introduction of terrestrial DTV in Japan was made by replacing analogue with digital broadcast content; in other words, DTV was not regarded as a new service under new rules or new regulations, but as the same service as analogue broadcasting, in spite of the fact that spectrum blocks were assigned anew to the terrestrial broadcasters for DTV on top of ones having been assigned previously to analogue broadcasting.[6] Thus, at the time of the transition to DTV, few changes were made in rules or regulations applying to the terrestrial broadcasting industry by the Japanese government. In particular, no new entry into DTV was admitted; it is stated by the government that new entries in the future may be allowed once the DTV transition is complete.

The terrestrial television industry in Japan has been enjoying a monopolistic status for decades. In urban areas, the government issued virtually no new licences for terrestrial broadcasting for the reason that there was no spectrum available. In rural areas, spectrum was available; however, there were only a few licences issued for decades presumably since the demand for television broadcasting was not sufficient to justify new entries. As a consequence of such a policy, the average rate of profits in the Japanese terrestrial television industry has been quite high relative to that of other industries, although there is a difference between urban and rural broadcasters. Broadcasters located in Tokyo are called 'key network stations,' supplying most of the broadcast content in Japan. (There are a few 'quasi-key network stations' in Osaka, the second largest city in Japan.) The profits accumulated by the key network and the quasi-key network stations have been invested for the improvement of broadcasting equipment and also for producing high-quality public-appealing content. It is expected that broadcasters located in urban areas can bear the costs needed for the DTV transition, whereas those located in rural areas may not be able to bear all such costs.

Analogue terrestrial broadcasters in Japan were not eager to introduce DTV in an attempt to maintain their monopolistic status. First, the decision for transition to DTV was made at the initiative of MIC, not by the industry itself. Second, the industry leaders preferred to minimize the impact of DTV transition; in particular, they wanted to maintain unchanged the environment and the structure of the terrestrial broadcasting industry. The government, MIC, made its policies for DTV transition mostly along the lines requested by the industry.

As a consequence, the impact of DTV transition is explained to the Japanese public mainly from a technological angle, not from an industry or regulatory one. Major benefits of DTV are said to be:

1. noise reduction with clearer images and sounds;
2. the realization of high-definition TV (HDTV, called 'Hi-Vision' in Japan);
3. spectrum saving; and
4. the possibility of interactive TV.

It is understood, however, that the benefits of DTV are not limited to technological ones; the greatest potential benefit of DTV should lie in the fact that, because DTV carries its content in a digital form, it is possible to process DTV content after they are broadcast by utilizing the power of computers and software. Consumers could enjoy content along with other valuable features. With analogue television, the possibilities of content processing were limited; recording at a low quality was the only possibility. For the benefit of all consumers, it is desirable to prepare an environment for content processing as soon as possible and to the greatest extent; needless to say, copyright issues need to be taken into consideration. One of the objectives of this chapter is to consider what policies should be taken in such a preparation.

The Internet today is, by far, the greatest means for transmitting digital information. It can move documents, pictures, music, and web pages smoothly in the form of digital information. Although the capacity of the Internet to transmit digital video images may still be limited, it is expected that, in the near future, video images will be distributed on the Internet as freely as web pages are distributed today. For this reason, the relationship between DTV and the Internet is one of the most important issues in the transition to DTV.

In short, DTV and the Internet may be competitive in some aspects but complementary in others. Consumers will benefit more when DTV and the Internet are offered together rather than as two separate services. This expectation is expressed by the term 'convergence of broadcasting and telecommunications'.

From the standpoint of consumers, it is desirable to be able to choose DTV or the Internet as a means of sending or acquiring information, regardless of whether it is a video programme or another type of information. From the standpoint of content producers and providers, it is also beneficial to be able to choose DTV or the Internet or both as a means of delivery. In an ideal situation, a consumer or a producer/provider in information transmission need only consider the means best fitting their needs. Note that, at present, we send an e-mail or obtain web pages on the Internet without paying attention as to whether the means of transmission is optical fibre, spectrum, or satellite.

Currently, broadcasting is the most economical means for transmitting information uniformly to a large number of receivers.[7] The Internet, on the other hand, is the most economical means for transmitting information from a particular point to another point on earth. Between these two, the efficiency of transmission depends upon factors such as the number of origins, the number of destinations, or the required speed and the accuracy for each piece of information. Thus, DTV and the Internet may compete with each other in such cases, but, needless to say, competition promotes the growth of both.

It is possible, however, that DTV and the Internet co-operate, rather than compete, with each other. For example, consider a piece of information to be sent to a small number of people living in a particular area. The best means for transmitting such information could be a combination of the two. One possibility is that the information would first be transmitted via the Internet to a community broadcasting station located in the area in which the intended receivers live; a local broadcaster can then disseminate the information through wireless radio or TV. Relying only on DTV or the Internet, in such a case, may be quite costly. It is one of the purposes of this chapter to consider policies that lead to a smooth combination of DTV and the Internet for efficient transmission of information.

PLATFORMS FOR EFFICIENT UTILIZATION OF DTV CONTENT

Restrictions on content utilization with DTV in Japan

DTV in Japan is under rigid restrictions on utilizing content by viewers. First, all DTV programmes are broadcast scrambled. In order for a viewer to have them descrambled, it is necessary to have an IC-card (called the B-CAS card) inserted in the proper slot in the receiver. B-CAS cards are issued by an organization, created and controlled by the terrestrial broadcasters in Japan, to manufacturers of receivers who have agreed to comply with the specifications required by the organization in producing receivers. In particular, a qualified receiver must satisfy a 'copy-once requirement'. In effect, a viewer can store or copy digital content only once; if the viewer creates a new copy, then the original is automatically deleted (hence, 'copy once').

Such viewing restriction and copy protection was implemented first by satellite broadcasters (the BS broadcasters) when DTV on BS was begun in 2000. (This is why the name B-CAS card is used.)[8] Terrestrial broadcasters have agreed to use the same card that is used for DTV on BS. No legislation in broadcasting gave this right to broadcasters except that MIC did not oppose the introduction of the B-CAS card to BS and the terrestrial DTV. MIC would justify this action (or inaction) on the ground that it is endowed with the administrative task of overseeing and regulating the conduct of broadcasters. Thus, DTV content is protected in Japan by the copyright law through the system of B-CAS cards.[9]

The introduction of the B-CAS card to terrestrial DTV means that it is now possible for a broadcaster to introduce pay TV, that is, to charge a subscription fee to watch a channel or a programme. At present, however, no commercial broadcaster in Japan has introduced, or intends to introduce, pay TV, since it is believed that free commercial TV is the most profitable way of broadcasting.[10,11] In the future, however, depending upon the speed of technological developments, new services such as 'TiVo' in the US (DVR service) may become so popular that it may be more profitable for a broadcaster to supply a portion of its content not as a free commercial TV but as pay TV.[12]

From the standpoint of copyright protection, it is understandable that terrestrial DTV broadcasters introduced such a stringent restriction on viewing DTV programmes. Copyright laws in Japan protect content producers including DTV broadcasters. In reality, however, it is likely that, without some effective means enforcing such protection, DTV content could be copied and distributed widely, as music content has been for years. Because of the introduction of the B-CAS card and the outright prohibition of copying from DTV programmes, content producers and broadcasters will be heavily protected; at the same time, however, this makes it impossible for a third party to process DTV content into a value-added product for consumers. In other words, if present-day protection of processing DTV content continues indefinitely into the future, we will lose an enormous benefit which may be obtained from processing such into values-added products and services for consumers.

The present state of DTV copyright protection in Japan may be illustrated by means of Figure 12.1, in which a tradeoff of the degree of copyright protection with the possibility of developing applications software for processing DTV content is shown. When protection becomes more stringent, then the possibility of developing applications software becomes less (and vice versa); this relationship is expressed in the diagram by a downward-sloping curve. The present state is expressed by the lower right-hand corner of the tradeoff, at which point the degree of copyright protection is maximized on the one hand, but the degree for the possibility of developing applications software is minimized on the other hand. A choice for society may be somewhere in between the extremes; it is shown as an optimal state in the diagram. Public policies for DTV transition should consider, in the long run, realizing such an optimal point.

Platforms for DTV applications

Applications software developed for DTV content, which can be installed on DTV receivers in the same way that computer software is installed on computers, can provide a great deal of satisfaction for consumers. The potential benefits of opening up the possibility of developing applications

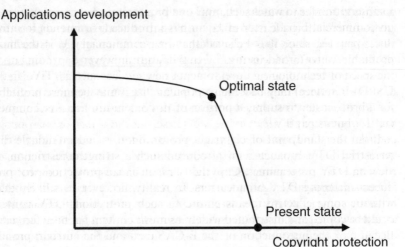

Figure 12.1   Tradeoff of copyright protection and applications development

software for DTV content are so great for both consumers and producers that it is impossible to spell out even a portion of them. The benefit from DVR services is but a small one.[13] The present state of DTV, in which we have a lot of content but no applications software, may be compared with the state of computers in earlier days, in which there were little applications software but a lot of analogue content in the form of documents and statistics printable on paper.

One condition for the development of applications software for DTV content is to prepare an environment for transactions of content with a copyright. Copyright laws in Japan protect owners of a basic copyright and the rights derived from it such as duplicating, modifying, distributing (including web distribution) content, and so on. Business codes in the form of established rules of conduct for selling and buying (a portion of) a copyright attached to DTV content, however, are yet to be formulated in Japan. Database and network systems, which can support smooth transactions of a copyright, seem to be only at the stage of designing or testing, at best. Thus, a great amount of work is left in legal, business, and technological arenas before realizing smooth transactions of digital content with a copyright.

In the remaining portion of this subsection, we shall consider a 'transaction system for DTV content'. Basically, it is an extension of the system for transacting goods and services, that is to say, the market mechanism. Goods and services are produced and sold freely by producers, and purchased freely by consumers, with a price attached to each object transacted. In extending the market mechanism to transactions of content, we need to pay attention to the differences between (ordinary) goods and services and 'content'. As widely known, the most important difference is that content, unlike ordinary goods or services, can be copied with or without modification; technologically, there is no limit to making copies or adding modification. This is the reason we say that the potential benefits from utilizing content are great, but at the same time, it is the reason that copyright laws introduce a variety of rights derived from the basic copyright. Thus, the degree of complexities in transactions of content, including DTV content, is far greater than that in transactions of ordinary goods and services; we cannot avoid dealing with such complexities if we seek the benefits of using content extensively. In short, the cost of transacting content is high. However, for almost all cases, transactions of digital content are conducted on contracts written electronically, not on paper; the cost can be saved, accordingly.[14] We need a framework supported by computer applications for managing transactions of content.

The following is a brief outline of a system for transactions of digital content, to be built on copyright laws. First of all, we need to understand that the object of a transaction here is not content itself, but a right (or rights) attached to it. For example, the producer of content initially possesses all the rights attached to it, and may wish to sell the right to make 10 copies with prescribed restrictions on using each of the 10 copies. Since there can be many 'rights' in relation to the content, and the value of each 'right' may depend on the status of the other 'rights', it is necessary that the status of the rights attached to the content be known to those interested in transacting one of them. This means that we need to create an information set which spells out the status of each right attached to the content and disclose this information to those interested in a transaction of such.

What is stated above may be realized first by creating, for content offered for transaction, a file containing the information which fully spells out the current status of the rights attached to the content; we call this file the 'descriptor'. Next, we need a database of the descriptors of content. This database should be disclosed to the public; each entry of the database should be administered so as to reflect transactions of all of the rights attached to content. For example, after the right to make 10 copies of content is sold to Mr A with restrictions in using each copy, suppose that the original producer of the content intends to sell to Mr B the right to make additional five copies of the same content. Mr B should pay for this not only to the original producer of the content, but also to Mr A, since the value of the 10 copies obtained by Mr A is decreased by the creation of the five additional copies by Mr B. These transactions between the producer,

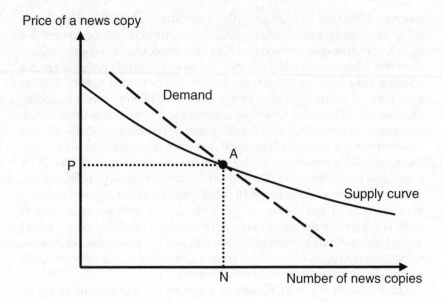

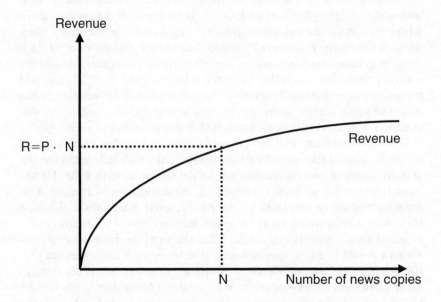

Figure 12.2a   An example of transaction of copies of DTV News

Mr A, and Mr B could be done privately without disclosing the information about the transactions. In order to realize smooth transactions of a large number of content by many participants, however, it is necessary to assemble all information about transactions into a database and disclose it to the public; otherwise, transactions cost would be prohibitively high. Observe that we do have such a system at present; the record of transactions of real estate is registered and disclosed; without such a system, real estate may not be transacted as smoothly as it is today. Although the size of a database is greater in transactions of content than of real estates, the reason that a database for transactions need to be disclosed is the same with content as with real estate.

The following is an example of a simple descriptor of DTV news content which may be offered for sale by a DTV broadcaster daily for, say, school teaching. Suppose that the broadcaster offers copies of daily news as represented by the supply curve in Figure 12.2a, in which the supply price of a copy of the news decreases, but the total revenue from selling it increases, as the number of copies sold increases. Consider a system in which each potential buyer of the news registers the maximum price at which he/she is willing to pay for a copy; this will form a demand schedule as represented by the demand curve in Figure 12.2a. Suppose that, when the offer is closed (say, two hours after the news was broadcast), an equilibrium point like A in Figure 12.2a is found with the given supply and the demand curves. Then, N copies of the news will be sold to those having registered at a price greater than or equal to the level P; the broadcaster will obtain the revenue $R = P \cdot N$. This process may be repeated; the second round of registration may, say, start one-hour after, and end one-day after, the first round was closed; the supply schedule in the second round should be lower than that in the first round. And so on.

By means of such a system, broadcasters can expect additional revenues from selling news copies, and school teachers can enrich their teaching materials by using a copy of the news. It is conceivable that a vendor specializing in the production of teaching materials from daily news participates in the formation of the demand schedule. Thus, a school teacher, on Monday morning after a weekend with a major M&A news, may find good teaching material waiting for him/her on the desktop of a computer.

All of such transactions must be done with computers and software which comply with an extended version of the copy-once restriction under the B-CAS card in the same way as DTV receivers and digital content recorders comply at present.

Figure 12.2b gives an example of entries in the descriptor of DTV news, which is put for sale by the broadcaster. The descriptor is stored in a database and is disclosed to the public. Entries in Item 1 of Figure 12.2b

Descriptor of Video Copy for Transaction

1. Video Characteristics

Title: Morning News   Class: General News

Broadcast by: XYZ Network, Inc.   at: Tokyo and other locations

on: 02/15/2005

time: 9:00–10:00AM

duration: 60 min

Commercials: not included

2. Supply-Price Schedule

Price per Copy (yen)	No. of Copies registered for sale
1.000	~500
700	~1 000
500	~2 000
310	**~5 000**
220	~10 000
160	~20 000
100	~50 000
80	~50 000

3. Current Registration for Purchasing a Copy

No. of Copies for Effective Purchase: **2513**

No. of Copies Delivered: **0**

Price Currently Bid: **310** yen

Closing Time of Current Offer: **Noon, 2/16/2005**

---

4. Conditions for Copy Utilization

   a. Must use equipment with B-CAS compliance

   b. Utilization Type: **A2**

      (1) May retain single copy for viewing;

         Duplicate copies not allowed.

      (2) May cut and use any portion(s) of video, but only one cut is
         allowed.

         Duplicating cut portion(s) not allowed.
         Cut portion(s) will be removed from the original video; the
         original video will be shortened by the (total) length of cut
         portion(s).

      (3) Any video to which cut portion(s) of the original video are pasted
         may not be duplicated; such shall be used under the copy-once
         restriction.

Figure 12.2b   Example of descriptor of news video for sale

summarize the characteristics of the news. A supply schedule is given in Item 2, which is a list of pairs of price and quantity of content copies; it shows that the broadcaster is willing to sell news copies in quantity, say, greater than 2000 and less than 5000, at the price of 310 yen per copy. This schedule corresponds to the supply curve in Figure 12.2a. Next, the potential buyers of the news, whose preferences are aggregated into the demand curve in Figure 12.2a, may register individually of the number of copies to be bought and the maximum price payable for them.

Item 3 in Figure 12.2b exhibits the current state of the supply and the demand as matched on the database by a computer. The example reads that, currently, at least 2513 copies of the news will be sold at 310 yen per copy or lower; the number of copies sold may increase and the price may decrease, if additional demand is registered by potential buyers. Delivery of copies will be done after the offer is closed; the broadcaster selling a copy of the news will send a 'key' to each buyer for working with a copy of the news (200 keys to a vendor buying 200 copies). This completes a round of the sale of the news.

Item 4 in Figure 12.2b outlines the conditions of using a copy of the news, as set out by the broadcaster. This example shows what may be called 'cut-and-paste once' restriction, which is an extension of the 'copy-once' restriction by one step. A user may cut any portion(s) of the news and paste them into another video only once; no more than one copy of any portion(s) of the copy of the news sold to the user can exist at any time. Further, each video to which portion(s) of the news are pasted should be subject to the copy-once restriction. Note that, although 'cut and paste once' is a slight relaxation of 'copy once', the potential benefits from it may be enormous, since, for example, a large number of school teachers can enrich their teaching materials by means of utilizing portion(s) of broadcast news.

Designing the descriptor of content and a database of descriptors is a work which should be done from engineering, economic, and legal expertise. In particular, copyright laws should be revised so as to accommodate transactions of the copyrights of DTV (and other) content. Further, construction of a system for transacting digital content should be done experimentally step by step. Naturally, a system for simple transactions of valuable (expensive) content should be constructed first; those for complicated transactions should come later. We look forward to seeing the development of a market mechanism for transactions of DTV and other digital content.[15]

There is one more point worth considering for developing applications software for DTV content; it is the status of broadcasters. As stated previously, for historical reasons, broadcasters in Japan have maintained a unique status legally and economically. The economic status of broadcasters is that of a monopoly in the supply of broadcast content; this status has been protected by the government by not allowing new entries into the broadcasting industry for the purported reason of the shortage of radio spectrum. Regarding the legal status of the broadcasters, NHK is a public entity under the NHK law, while commercial broadcasters are private profit-seeking corporations. In fact, however, because of certain regulations, even commercial broadcasters have considerable obligations to the public. We may state that these obligations are, in effect, imposed on the broadcasters in exchange for the privilege of using the radio spectrum for broadcasting.

Because of the economic and legal status of the broadcasters, as stated above, there is little incentive to let DTV content be utilized with applications software for the benefit of consumers. In other words, the monopoly benefits enjoyed by the broadcasters, at present, seem to exceed the potential profits obtainable by supplying DTV content for applications software. There are two ways to alter this: one is to increase the potential profit of supplying DTV content for applications software, which was discussed in the first part of this subsection, and the other is to decrease the monopoly profit of the broadcasters. This will be discussed in the following section in relation to the competition and the co-ordination of DTV with the Internet.

There may still be yet another way to force broadcasters to let DTV content be utilized with applications software; that is by means of direct governmental regulations. It is conceivable for MIC to introduce, if step by step, 'disclosure obligation' of DTV content on the broadcasters. The first choice of disclosure may be information supplied by the government itself such as an interview with the Prime Minister or a video record of a session of the Parliament. The second choice may be a news item, for which copyright protection may not be important. Other choices include educational, medical, scientific, or welfare-related content. Prices may be attached to the supply of such content subject to governmental regulations. Such a regulatory solution, however, should be considered for a short-term purpose only, since such will always bring distortions and inefficiencies.

COMPETITION AND CO-ORDINATION OF DTV AND THE INTERNET

Vertical Structure of DTV and the Internet

In order to realize the benefits of competition and co-ordination between DTV and the Internet, it is necessary to introduce a business environment in which there is fair competition at a level-playing field. We start with an understanding of the present situation in terms of a vertical structure of the communications industry including DTV, the Internet, telephony, and others.

Figure 12.3 outlines a vertical structure in the communications industry. The top row lists communications services classified traditionally: telephony, the Internet, cable TV, and broadcasting. In the left column, from top to bottom, we list services classified in vertical layers: content, networking, (physical) media, and infrastructure (equipment, structures and spaces). Thus, when e-mail or web pages are transmitted on the Internet, they are first put into the form of IP-packets for networking, and then transmitted via cable such as twisted copper pairs, coaxial cables or optical fibres. Further, those cables are laid in tunnels, tubes, or between poles, which are constructed in publicly-owned spaces. In the case of broadcasting, the layer structure is simpler; after broadcast content is created, broadcasters (key network stations) transmit it to local broadcasters, where content is modulated and put on the radio spectrum. Spectrum with content emanates from broadcasting antennas to receivers' antennas; the resource devoted to this is the terrestrial spectrum space.

It is noted that the price of information transmission which a consumer pays can be divided into components corresponding to the service of these

Services	Telephony	Internet	Cable TV	Broadcast	Mode of Supply
Contents	Contents of telephone and fax	E-mails, Web	Broadcast contents		
Networking	Voice transmission	IP-Packet transmission	Cable transmission	Broadcasting	Competitive
Media	Electric current, Optical rays			Spectrum	
	Twisted and coaxial cables, Optical fibres			(Antennas)	
Equipment structures and spaces	Tunnels, tubes, poles, etc.			Terrestrial spectrum spaces	Monopolized
	Terrestrial (physical) spaces				
Infrastructure	Wired			Wireless	

Figure 12.3   The layers structure of the services for information transmission

layers. This is similar to the price of bread, which can be divided into payments to a flour producer, a mill operator, and a farmer producing the wheat. Thus, the vertical layers of the communications industry are nothing but a division of labour viewed vertically. For historical reasons, however, the layer's structure has not received much attention by the communications industries. Telephone operators and broadcasters were born as vertically integrated entities; accordingly, vertical division of activities into layers was not interesting. Once digital technology was introduced into the communications industry, however, the division of activities into vertical layers became interesting and important, since it brought the possibility of vertical division of labour for increasing the overall efficiency, typically seen in the computer industry as the division into hardware and software.

The introduction of DTV in the broadcast industry brings, from this standpoint, the possibility of a new vertical division of labour. The potential benefits from the competition and co-ordination between DTV and the Internet are a consequence.

Monopoly in the infrastructure

We observe that the benefits of competition and co-ordination in the communications industry arise with activities competing (and thus substituting) with each other within a single layer. A classical example is the shift from twisted copper pair to optical fibres in the layer of transmission media. The shift of the means of telephony from traditional voice transmission to new IP-packet transmission is taking place at the layer of networking due to the efficiency of packet transmission over non-packet transmission. Another example is a change in the distribution of broadcast content from traditional transmission by means of terrestrial spectrum into cable transmission; the reason for this change was the efficiency of the combination of cable with satellite transmission as opposed to transmission relying only on terrestrial spectrum. Most long-distance transmission of broadcast content in Japan at present, however, uses optical fibres.

These examples show that, as a particular service in a layer becomes more economical, the substitution of new technology for old takes place. This is the basis for the benefits of technological progress to be enjoyed at the level of consumers.

For this reason, we can state that, in the digital world, it is best to promote competition layer-wise. By removing barriers to mutual entries within each layer, we can expect that new technology can be smoothly deployed; in addition, such will encourage further technological progress, ultimately increasing benefits to consumers.

In the following, we shall concentrate on the single most important factor impeding the promotion of layer-wise competition in the communications industry: the monopoly in the infrastructure layer. In Figure 12.3, the layers are divided into two main groups: competitive and monopolized, as shown in the right-hand column. The double solid lines in the diagram indicate the boundaries between the two groups.

We first note that, in the communications industry, an operator must directly or indirectly use some publicly owned space. In the case of wired transmission, structures for communication such as tunnels, tubes, and poles are constructed by using physical space, which is land, underground, or underwater. The value of a structure is composed of the cost of the structure itself and the value of the underlying space (for example, the value of the land) on which it is constructed. When it comes to wireless communication, the notion of infrastructure is not so clear, since an unseen entity, radio spectrum, is used as a means to transmit information. We can then consider an underlying space to be the terrestrial spectrum space (as opposed to the physical space), on which the transmission takes place. There is no structure such as a tunnel or a cable used for wireless communication; hence the cost of infrastructure for wireless communication arises almost exclusively from the cost of spectrum spaces, which are a scarce resource today.

For historical reasons, the legal and economic basis of the supply of the infrastructure layer is not clearly established, nor is it at a level-playing field with a competitive price. In the case of wired communication, the NTT Corporation supplies a large portion of the communications infrastructure, which was given to it at the time of its privatization. There may be an accounting of its infrastructural equipment and underlying spaces, but it is only nominal and departs from the real economic value. For wireless communication in Japan, the right to use radio spectrum is assigned by MIC to users without charging according to real economic values. In the case of the broadcasting industry, in addition, the supply of radio spectrum to certain broadcasters has generated their monopolistic power.

In short, the way in which the services of the infrastructural layer are supplied, in Japan, is far from being competitive or with free entry; vertically-integrated operators such as broadcasters or NTT may freely charge for such infrastructural elements enjoying monopolistic profits or an advantage of internal cross-subsidization in upper-layer competition. In order to promote fair competition on a level-playing field, we need to deal in some way with monopoly in the infrastructure layer.

Policies for fair competition at a level-playing field

In this subsection, a proposal will be made for a system by means of which the evils of monopoly in the infrastructure layer can be minimized by appropriate governmental regulations. The basic idea for this is to regulate the supply in the monopoly layer so that the supply be made as if it were a competitive supply.

In order to do this, we must first distinguish monopolistic services from competitive ones as indicated by the double solid lines in Figure 12.3. Let us define, for each communications service supplied to consumers (final users), the 'monopoly-front service' as that service located at the highest layer within the monopolized group. In Figure 12.3, the monopoly front for telephony, the Internet, and cable TV are the services supplied in the layer with tunnels, tubes, poles, and so on. For broadcasting, the monopoly front is the service of (terrestrial) spectrum. Thus, the level of the monopoly front may not agree among different services. The determination of the location of the monopoly front should be done by the government, considering the degree of monopolistic power of the service in question. In short, when new entry is possible, the service should not be included in the monopolized group. Hence, in the long run, the location of the monopoly front may change dependent on the possibility of new entries.

The basic idea of introducing the concept of a monopoly front is to regulate the supply of the services located on or below it at the front level so that the monopolized group functions as if it were a competitive group. This can be done in the following ways by means of governmental regulations.

First, the government should regulate each operator so that monopolized services, such as communications infrastructure, be vertically separated from competitive ones, regardless whether the infrastructure is wired or wireless. The separation may be structural in the sense that a vertically integrated operator is divided into two operators, or it may be of accounting without actually dividing the operator. In either case, there should be no regulation on competitive activities. In contrast to this, the supply of monopolized activities, in particular the supply of services in the monopoly front, should be regulated in the following way.

Consider the short-run behaviour of the monopolistic operator in supplying a monopoly-front service.

First of all, the supply of a monopoly-front service should be open to all purchasers without discrimination. If the monopolistic operator, supplying the monopoly-front service, is structurally separated from competitive operators, then fair transactions in the market of the monopoly-front service can exist. If the monopolistic operator is separated from competitive activities in accounting only, this requirement implies that the operator should publish the internal price in which the monopoly-front service is sold, and offer such a service to outside purchasers at a price equal to that used for internal transactions. We call this requirement 'no discrimination requirement'.

The second requirement in the supply of a monopoly-front service is that the monopolistic operator must act as a price taker. This means that the monopolistic operator first determines the quantity of the monopoly-front service to be supplied for a time period (for example, a year), and then sell it at a price with which the demand for it is equal to its (fixed) supply. The monopolistic operator is not allowed to withhold a portion of the monopoly-front service in order to raise the price; this means that the operator is prohibited from charging a monopolistic price. We call this requirement 'price-taker requirement'. See Figure 12.4.

It is clear that the two requirements imposed on a monopolistic operator by the government enforce a monopoly-front service to be supplied at a price at which the demand for, and the supply of, the service is equal, that is, at a competitive price. Such a price of a monopoly-front service includes all the costs incurred to its layer and to the layers lower than it. The price of a

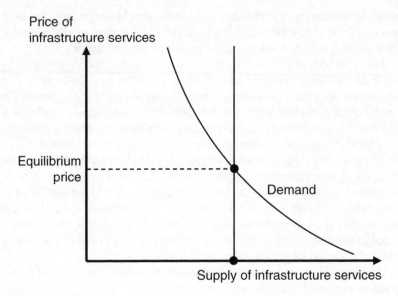

Figure 12.4   Equilibrium price of infrastructure service

monopoly-front service will be high in urban areas in which the demand is high, while the price in rural areas may not be as high. In short, the price of a monopoly-front service shows the value of the infrastructure supporting the service.

If the infrastructure in the communications industry is supplied competitively in the sense described above, then the evils of monopoly such as internal cross-subsidization are effectively removed and, as a consequence, operators in the competitive layers are assured of a level-playing field. Most of the difficulties and the complexities in the communications industry arise from the fact that every activity in it must use, directly or indirectly, the service of some infrastructure (including space), which cannot be supplied competitively without governmental regulations. Thus, the requirements imposed on the monopolistic operator make it possible for competitive operators in the communications industry to behave as if they were in a competitive environment. In short, the requirements are a way to transform the communications industry, which cannot operate competitively without governmental regulations, into one similar to other industries in which no monopolistic element exists. Figure 12.5 summarizes the situation of the communications industry, including DTV and the Internet, after separating it into the competitive and monopolistic layers.

The remaining portion of this subsection will be devoted to the discussion of the long-run behaviour of a monopolistic operator. It is how the

Services	Economic units		Mode of supply
Final demand	Consumers, Firms, Governments, Public entities, etc		
Content supply	Content suppliers (newspapers, publishers, producers of music and video contents, news agencies, advertising agencies, etc.)	Web, E-mails, and other data	Competitive
Information transmission	Network-service providers (broadcast, telephone, access, backbone, Internet, etc.)		
Infrastructure	Supply of infrastructure for information transmission (wired, wireless)		Monopolized

Figure 12.5   The structure of IT industries under vertical separation

monopolistic operator should construct and maintain the infrastructure under his control. The short-run behaviour of the monopolistic operator is to simulate the working of a (short-run) service market. In the same way, the behaviour of the monopolistic operator with regard to the construction and the management of the infrastructure, in the long run, should be to simulate the working of the competitive capital market. Thus, the monopolistic operator should invest in constructing additional communications infrastructure whenever the expected rate of return from it is greater than the expected interest rate to be paid on the fund needed for investment. The monopolistic operator is *prohibited* from maximizing the rate of return from investment, exactly for the same reasons as the operator is prohibited from maximizing profits (by means of imposing a monopolistic price). In this way, the evils of monopoly will be removed in the supply of the communications infrastructure in both the long and the short run.

A form of organization fitting the behaviour required of a monopolistic operator may be a 'public corporation', for which the main objective is not maximization of profit or rate of return, but simulation of competitive behaviour in the short run and in the long run. Competition may be introduced among such public corporations, but they should be evaluated in terms of how they simulate competitive markets, not on how well they make money. Observe that it is possible for the government to encourage investment in a particular communications infrastructure, if so chosen, by means of a subsidy on interest payments given to the monopolistic operator managing it.

The following is a list of policy recommendations for promoting competition and co-ordination between the Internet and DTV. First, the supply of the infrastructure for data transmission on the Internet should be reformed to satisfy the monopoly-front and the price-taker requirements. For example, if a telephone operator supplies optical fibres to Internet operators, then his activities should be divided, at least in accounting, into competitive and monopolistic ones. Further, the supply of structures such as tunnels, conduits, or pole spaces (in case fibres can be constructed freely so that the supply of structures is at the monopoly front) should follow the price-taker requirement.

Second, the supply of broadcasting services should be reformed to satisfy the monopoly-front and the price-taker requirements. Consider a case in which the supply of spectrum for broadcasting is at the monopoly front. Then, in order to satisfy the two conditions, it is necessary to supply spectrum competitively to broadcasters without discrimination between incumbents and newcomers. A way to do this is to introduce competitive lease of spectrum to broadcasters possibly with auctions on lease prices at the initial and renewal assignments of it.[16]

Once such policies are implemented, then the distribution of DTV content can be made competitive and on a level-playing field. In particular, a broadcaster can choose and combine both wireless and wireline means to supply DTV content. An Internet-service provider can also work wireless and wireline. Further, the content provider (producer) can choose and combine the service of a broadcaster and that of an Internet operator. Competitive prices will be formed for alternative means of transmitting digital content; technological advances, not regulatory nor monopolistic factors, will be the main determinant of change in such a competitive environment. Thus, the activities of broadcasters and Internet operators will be directed by technological advances; this in turn will encourage technological advances. In this way, we can expect that the welfare of consumers is increased through co-ordination and competition between DTV and the Internet.

NOTES

1. For instance, services such as 'TiVo' in the US let viewers watch programmes at a time and in an order chosen by them and greatly increase the benefits of TV to viewers (http://www.tivo.com).
2. Needless to say, one should take into account copyright matters. See the third section.
3. Recall that when television or the Internet first became available to us, we did not foresee the enormous impact that the new systems would bring.
4. See MIC (2004a).
5. See NHK (2004). Per-person time devoted to watching TV is approximately 1 hour per day (MIC, 2004).
6. It is noted that, because of the crowded spectrum blocks assigned for broadcasting in Japan, a process of reshuffling them for analogue broadcasting has been under way since 2003 in order to open up spectrum blocks for digital broadcasting.
7. Note, however, that the price of using spectrum for terrestrial broadcasting is currently set to be zero by the government; that is, MIC issues a broadcasting licence free of charge (Oniki, 2005). For this reason, broadcasting, as a means to transmit information uniformly to a large number of receivers, may appear distortedly more economical than the Internet.
8. CS broadcasters use the card called 'C-CAS card', which functions in the same way as the B-CAS card.
9. In the US, the Federal Communications Commission (FCC) has a law for protecting DTV content by attaching to content a 'broadcast flag', which is a name given to technology, hardware or software, which makes it possible for content providers (broadcasters) to prevent DTV programmes from being copied and distributed on, say, the Internet. DTV receivers supplied on or after 1 July 2005, must comply with the broadcast flag requirement. This restriction, unlike the one in Japan, does not require that DTV content be scrambled, and it allows a broadcaster to adopt a technology from those certified by FCC for flagging its content. Further, whether to flag or not is up to the broadcaster. FCC explains that the introduction of a broadcast flag is to foster rapid transition from analogue television to DTV by encouraging broadcasters to supply DTV content without fearing that they are copied and distributed illegally on, say, the Internet. See FCC (2002, 2003, 2004).
10. The revenue received by the TV broadcasters for advertisement occupies the largest share of one-third in the total advertisement revenue in Japan (See, for example, MIC, 2004 a).
11. It is noted that the public television broadcaster in Japan (NHK), a nonprofit organization under a special law, charges a subscription fee from each owner of a TV receiver. The introduction of the B-CAS card has made it possible for NHK to collect this fee more effectively (the ratio of the number of viewers paying the fee to the total TV-receiver owners has been around 80 per cent in recent years; there is no penalty for declining to pay), and also to collect it based not on the possession of a TV receiver but on the choice of watching NHK programmes.
12. DVR service is called 'server-type TV' in Japan. It is a way to watch a TV programme by first storing it in a storage device such as a hard-disk, and then actually watching it according to the time chosen by the viewer. In effect, it becomes possible for a viewer to skip advertisement portions (commercials) freely; hence, with this device, free commercial TV may no longer be a profit-making method of TV broadcasting.
13. One immediate example is application to school teaching. TV content such as news or news analysis may be used for producing teaching materials with the aid of applications software. Teachers may teach in class with a video which includes a fresh news his/her students watched just a few hours before.
14. For the basics of copyright economics, see Chapter 10 of this book written by K. Domon and E. Joo and, more broadly, Chapters 1–6 of Landes and Posner (2003).
15. We also note that the absence of a market mechanism for transactions of DTV content provides a strong incentive for breaking copyright protection illegally, since there always

exist a number, if small, of parties who wish to obtain a copy of content even at an extremely high price.

16. It may be necessary to protect the investment made by an incumbent broadcaster at a renewal auction. See Oniki (2002).

REFERENCES

Federal Communications Commission (FCC) (2002) 'Notice of proposed rule-making in the matter of digital broadcast copy protection,' MB Docket No. 02–230, FCC 02–231, adopted: 8 August 2002 and released: 9 August 2002, (http://hraunfoss.fcc.gov/edocs_public/attachmatch/FCC-02-231A1.pdf).

Federal Communications Commission (FCC) (2003) 'Report and order and further notice of proposed rulemaking in the matter of digital broadcast content protection,' MB Docket 02–230, FCC 03–273, adopted: 4 November 2003 and released: 4 November 2003, (http://hraunfoss.fcc.gov/edocs_public/attach-match/FCC-03-273A1.pdf).

Federal Communications Commission (FCC) (2004) 'Order in the matter of digital output protection technology and recording method certifications,' MB Docket No. 04–55, etc., FCC 04–193, adopted: 4 August 2004 and released: 12 August 2004, (http://hraunfoss.fcc.gov/edocs_public/attachmatch/FCC-04-193A1.pdf).

Landes, William M. and Richard A. Posner (2003) *The Economic Structure of Intellectual Property Law*, Cambridge, MA: The Belknap Press of Harvard University Press.

MIC (2004a) 'Revenue and expenditures of broadcast operators (in Japanese), MIC, (http://www.johotsusintokei.soumu.go.jp/field/housou03.html).

MIC (2004b) 'Average time spent for watching radio and television (in Japanese), MIC, (http://www.johotsusintokei.soumu.go.jp/field/housou06.html).

NHK (2004) 'Survey of television watching by individuals (in Japanese),' Broadcasting Culture Research Institute, (http://www.nhk.or.jp/bunken/index-e2.html).

Oniki, Hajime (2002) 'Modified lease auction and relocation – proposal of a new system for efficient allocation of radio-spectrum resources,' *ITME Discussion Paper*, 108, Information Technology and the Market Economy Project, Faculty of Economics, University of Tokyo, April (http://www.osaka-gu.ac.jp/php/oniki/noframe/eng/publication/200208.html).

Oniki, Hajime (2005) 'Spectrum Policy,' in R. Taplin and M. Wakui (eds) *Japanese Telecommunications Market and Policy in Transition*, London: Routledge.

Shapiro, Carl and Hal R. Varian (1999) *Information Rules*, Boston, MA: Harvard Business School Press.

13. Economies of scale, scope and vertical integration in the provision of digital broadcasting in Japan

Hitoshi Mitomo and Yasutaka Ueda

INTRODUCTION

Recent developments in information and communication technology have enabled the emergence of a new system of broadcasting. The Japanese broadcasting industry is experiencing a transition to digital terrestrial broadcasting. In December 2003, it began in three metropolitan areas of Japan: Tokyo, Nagoya and Osaka. In 2005, local key stations in regional core cities including Sendai, Mito, Yokohama, Shizuoka, Toyama, Kyoto and Kobe launched digital services. By 2006, all local terrestrial broadcast stations must switch to digital in compliance with requirements from the Ministry of Public Management, Home Affairs, Posts and Telecommunications (currently, Ministry of Internal Affairs and Communications). Analogue broadcasting is scheduled to terminate by 2011. By that time, digital coverage is expected to be nationwide and digital receiver ownership will reach approximately 85 per cent. Insofar as broadcasting is regarded – reasonably or not – as a 'universal service', programming must be able to reach every individual.

In addition, with broadband technologies, TV programmes can be transmitted through telecommunications networks. This suggests integration between broadcasting and telecommunication in the future. In fact, provision of TV programmes through broadband networks began in 2004. Telecom firms may be expected to supplement or even take over some traditional broadcasting functions.

Implementation of digital terrestrial service was estimated to cost around 6.3 billion Japanese yen per commercial station.[1] In addition to the technological advancement, the cost burden due to digitization may trigger reorganization of the broadcasting industry. For local commercial broadcasting stations especially, the amount seems to be not at an affordable level. With the advent of digitalized terrestrial broadcasting, socio-economic conditions for the Japanese broadcasting industry have become increasingly severe due to the large outlay of money required for technological implementation.

The commercial broadcasting industry has been characterized by an extraordinarily large number of stations: the total includes five key network stations, 109 subsidiary local stations and 13 independent local stations. Each station, whether key network, subsidiary or independent, is involved in both programme production and programme transmission. Each subsidiary local station televises programmes supplied by the key network station in addition to its own programming.

Since not all stations will be sustainable after nationwide digitization, some degree of reorganization of the industry will be necessary. There will be two possibilities for recognizing the industry: to extend the area covered by one station, or to separate functions or services each firm has for realizing the subsequent full-scale functional or service integration. The former is associated with scale economies and the latter with economies of scope or vertical integration.

This chapter aims to examine whether these economic effects apply to local Japanese broadcasting stations. An econometric approach tests the existence of economic effects. If the empirical evidence shows the existence of scale economies, then extending the area coverage will be more efficient for a local station. If economies of scope are present, separate operations for the services which a station provides will result in improved efficiency. Economies of vertical integration are necessary for a station to have vertically integrated functions such as a combination of programme production and transmission. After presenting empirical evidence, which suggests such economies do not in fact exist, we shall propose a vertical separation between the industry's programme production and transmission functions.

So far, Mitomo and Ueda have published several papers concerning an econometric analysis of broadcasting (Mitomo, 2000; Mitomo and Ueda, 2001, 2003; Ueda and Mitomo, 2002, 2003a, b and c; Ueda *et al.* 2004). This chapter contains findings from those papers.

BROADCASTING AND THE ECONOMIES OF SCALE, SCOPE AND VERTICAL INTEGRATION

Economies of Scale

A commercial local broadcasting station in Japan is licensed to broadcast in a local area; however, there is an ongoing debate over whether or not to consolidate the industry by relaxing the principle of abatement of mass media concentration. In other words, in response to the development of digitalization, local broadcasting stations are reconsidering their business efficiency by opting for large-scale operations.

For the broadcasting industry to pursue an economy of scale, stations could either increase their advertising revenue by providing high-quality programmes (expansion of operational scale) or increase their share of audience households by expanding the broadcasting area. Increasing the advertising revenue requires programming designed to attract larger audiences, which seems to be costly. The expansion of operational scale can be achieved through merging with stations in neighbouring prefectures. Consequently, the same content could be simultaneously distributed to a larger number of households. While there is a concern that spatial expansion (expansion of the broadcasting area) may result in a loss of localism, on the other hand, it would allow the elimination of regional differences in information and entertainment access for those locals when there is a lack of such access due to their limited number of viewing channels.

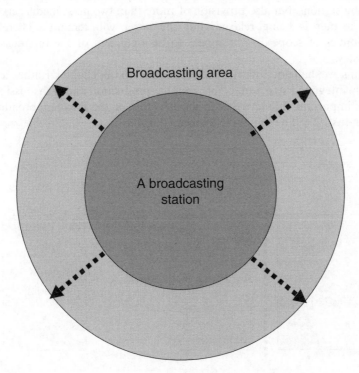

Figure 13.1   Economies of scale by expanding the broadcasting area

For spatial expansion, with the consolidation of the broadcasting stations to be a measure to improve operational efficiency, it would be necessary for the industry to show economy of scale. The details of an empirical test are discussed in the next section.

Economies of Scope

Economies of scope represent positive economic effects for multiproduct production. These economies exist if there is a cost advantage when products are supplied by a single company rather than when each product is supplied by a separate company. They are associated with the existence of a common cost. If a broadcasting station provides, in addition to TV programmes, other services such as radio programmes, newspapers, and so on, horizontally integrated production may show a certain economic effect.

However, supplying multiple services by one broadcasting station is strictly regulated by the principle of abatement of mass media concentration. The principle prohibits not only reciprocal ownership of broadcasting stations, but also provision of more than two mass media services. The former is being relaxed, but the latter will remain[2]. Therefore, economies of scope do not appear to be a concern of the broadcasting industry.

Thus, we shall not deal with such economies in this chapter. Rather, a vertically integrated structure of programme production and transmission, a more important characteristic in the structure of the Japanese broadcasting industry, will be subject to investigation.

A broadcasting station

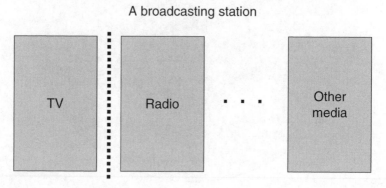

Figure 13.2   Economies of scope

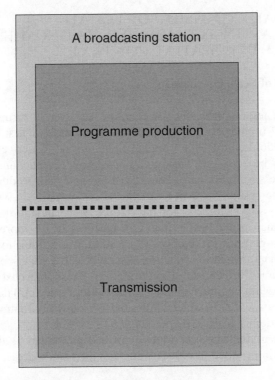

Figure 13.3   Economies of vertical integration

Economies of Vertical Integration

The dominance of the broadcasting industry in Japan derives from its own-ership of two key functions: transmission and content production (includ-ing the packaging of content into a distributable form). This vertically integrated structure gives it an exclusive and privileged status. Indeed, own-ership of transmission facilities is sufficiently costly to deter entry into the market. In addition, the scarcity of spectrum requires government alloca-tion of broadcasting rights, which further contributes to the privileged status of Japanese broadcast stations.

In an age of digital broadcasting, we may anticipate a value chain in which transmission function is vertically separated, because content providers will have new options for distribution such as the Internet and mobile communications to supplement existing terrestrial, satellite and CATV broadcasting. An empirical test is shown in the fourth section.

REGIONAL EXPANSION OF LOCAL BROADCASTING COMPANIES – ECONOMIES OF SCALE IN A SPATIAL CONTEXT

Formulation of Scale Economies

This section tests empirically the economy of scale in terms of spatial expansion of broadcasting stations. For this purpose, we first tried to formulate a hedonic model for broadcasting stations to derive the output explained by broadcasting area, area population, revenue, unit advertisement charge, total hours of broadcasting and self-production ratio. However, the results were not satisfactory. Alternatively, we took the broadcasting area as a proxy for the output. Since all stations in an area have the same output, while, it is not appropriate for representing the output of programme production of a station, it is suitable for representing that of the transmission function. A broadcasting area is multiplied by the total hours of broadcasting by the station. We adopt a cost function of the transcendental logarithmic form (hereinafter, 'translog') developed by Christensen *et al.* (1973), which allows flexible substitution among inputs. The formulation is given in Appendix A.

The index of the scale economies can be represented as the percentage increase in production cost (C) due to 1 per cent increase in production (y):

$$SE \text{ (Scale Economies)} = \frac{\partial \ln C}{\partial \ln y} \tag{13.1}$$

Scale economies exist in the neighbourhood of the average of samples if SE is less than unity.

Input Data

Price of labour, capital and other input production factors $(W_L, W_K$ and $W_o)$ are regarded as production factors. The price of labour is calculated by dividing the annual real personnel cost by the number of term-end total employees. The price of capital is derived from the Jorgenson-type user cost (excluding the exemption of investment tax):

$$UCC_{it} = \frac{1 - U_t Z_{it}}{1 - U_t} (\rho_t + \delta_i) q_{it} \tag{13.2}$$

where UCC_{it}:   the user cost of the capital stock i in the period t

ρ_t: The real return of investment in the period t

δ_i: The depreciation rate of the capital stock i

U_t: The corporate tax rate in the period t

Z_{it}:   The discounted present value of the future depreciation of the present unit investment of capital i

q_{it}:   The price of capital goods i in the period t.

Other inputs are aggregated and taken as a numeraire. Thus, the price is assumed to be a unity based on the premise that all broadcasting stations are faced with the same unit cost and that the cost represents the corresponding input. Other inputs include programmeme costs (such as performance fees, scriptwriting, and purchase of programmes), usage charges for transmission circuits, advertisement fees, and network guarantee fees. The personnel cost shares were around 25 per cent, which suggested that the broadcasting industry is a software industry, rather than a facility industry by producing content with input by human resources. The costs regarding fixed capital such as depreciation costs, running and maintenance cost is at most 10 per cent. As for local stations especially, most facilities were constructed when they began service and the remaining asset value should be low due to the depreciation over a number of years. Operating costs, including sales costs and sales management costs, are about 20 per cent while programme production costs are around 45 per cent varying according to the self-production ratio.

The data were collected for FY2000 from 37 of 109 local broadcasters, who submit financial statements to the Ministry of Finance. We excluded five key stations and 13 independent UHF broadcasters because, due to their location in metropolitan areas, their self-production ratio was extremely high. Deviations from the sample averages were used as input data for independent variables in the translog cost function.

The Results and Their Implications

The estimated parameters are all significant and satisfy the constraints on the parameters such as symmetry, homogeneity, monotonicity and concavity conditions. Based on the estimated cost function, the index of scale economies is calculated as $SE = 0.49$, which is less than unity and shows the presence of scale economies. (See details in Appendix C.)

The results suggest that a broadcaster with a wider coverage area has an advantage in efficient operation, and this may support a relaxing of the principle of abatement of mass media concentration. Commercial broadcasting is operated from commercial revenue, and to obtain more revenue, it is necessary to sell time frames for advertisements at a higher price. The price varies according to the volume of the audience, so that broadcasters can obtain more revenue if the area of coverage is wider.

Note that this does not necessarily represent individual cases but merely is a general tendency derived from a macroscopic view that stations with a wider area have a cost advantage. Distribution and density of population also influences the existence of scale economies. Furthermore, costs depend on geographical conditions. If an area is flat, the costs will be low. The Kanto, Kinki and Chukyo areas are typical examples. However, in mountainous areas or on islands, the costs increase as the area expands, but is relatively low compared to media as CATV.

POSSIBILITY OF SEPARATION BETWEEN PROGRAMME PRODUCTION AND TRANSMISSION – ECONOMIES OF VERTICAL INTEGRATION[3]

Formulation of Vertical Economies

With economies of vertical integration, the concept of multiproduct cost economies (economies of scope) should be modified to the case of production at vertically integrated stages (Kaserman and Mayo, 1991).

We test for the presence of vertical economies by estimating two cost functions. The first is the total cost function of existing broadcast stations in which both programme production and transmission are integrated. The second is the cost function of programme production, which is assumed to be independent of transmission and estimated separately. These cost functions can be denoted as:

$$C = C(Y; W_p; W_l, W_k, W_o) \tag{13.3}$$

$$C_p = C_p(X; W_l, W_k, W_o) \tag{13.4}$$

where C: The total cost of broadcasting

C_p: The cost of programme production
X: Output of programme production
Y: Output of broadcasting
W_p: Price of the input from programme production
W_l: Price of labour input
W_k: Price of capital input
W_o: Price of other inputs.

According to Hayashi *et al.* (1997), the index of vertical economies (*VE*) is defined as

$$VE \equiv 1 - [(\partial \ln C / \partial \ln Y) / (\partial \ln C_p / \partial \ln X)] \qquad (13.5)$$

A positive *VE* implies advantages to vertical integration. The numerator in the brackets is the marginal cost of broadcasting, which denotes the change in total broadcasting costs derived from an additional change in output. The denominator is the marginal cost of programme production. If *VE* is positive, vertical economies exist. If programme production is separable, an additional output causes the total cost of broadcasting to rise by more than the cost of programme production.

For an empirical analysis, we adopted cost functions of the translog form. In this case, scale economies of programme production and broadcasting are given by the coefficients of the first-order terms of X and Y, respectively. Details of the model specifications are given in Appendix B.

Input Data

Data were collected from 34 of 109 privately owned local broadcast stations, all of which submitted financial statements for fiscal years 1998, 1999 and 2000 to the Ministry of Finance. The results of the empirical test heavily depend on the data settings, with definition of the output of broadcasting and programme production particularly essential. In the analysis, we use the total revenue from broadcasting as the output of the former, and the revenue from locally produced programmes as a proxy for the latter.

Output, sales and revenues from broadcasting are ambiguous. Here, we classify them into four categories according to the source of programming and the business entities involved. For a local station, revenue will come from programming, direct transfers, programme production and provision of local programmes.

1. Programming: A local station provides a scheduled time to distribute programmes produced by its network key station and receives spot revenue from nationwide sponsors. The local station serves as a branch for the key network station.
2. Direct transfer: A local station serves only as a transmission medium to transfer programmes to viewers. Content produced by the key network station is distributed over the network without being processed by the local station.

3. Programme production: A local station produces local content (for example, local news) and distributes it locally. This yields revenues from local sponsors.
4. Provision of local programmes: A local station provides local content to its key network station and receives payment for the service.

We regard the sum of revenues in the four categories as the output from broadcasting, and the sum of (1), (3) and (4) as the output from programme production.

As in the case of scale economies, the production factor includes the unit costs such as labour, capital, and programme production. In the broadcasting business, unavoidable costs include overheads, programme advertising fees, and network guarantee fees. Because of the difficulties of obtaining data, costs (prices) associated with inputs other than labour, capital and programme production are categorized into 'other miscellaneous costs'.

Input factor prices are calculated in such a way as follows:

W_l:　Price of labour input = personnel cost/term-end total employees
W_k:　Price of capital input = capital cost/term-end value of tangible fixed assets
W_p:　Price of input from programme production = programme production cost/broadcasting hours of self-produced programmes.

The data used for this model were collected from *TV Research Weekly* for 34 local privately owned broadcast stations from 1998 to 2000.

The Results and their Implications

The results from an econometric evaluation of the degree of vertical economy among Japanese local broadcast stations, can be summarized as follows:

(a) Estimation of vertical economies (VE) is slightly negative, -0.0691.
(b) In detail, 20 of the 34 stations show negative vertical economies for the three years, another 10 were positive and the remaining four were a mixture of positive and negative.
(c) This test suggests that vertical integration does not imply an advantage (or disadvantage) with respect to cost savings for most local broadcast stations in Japan.
(d) Both broadcasting and programme production show economies of scale.
(e) Overall, major local stations tend to show strong negative vertical

economies. These stations release more self-produced programmes and, since the scale economies in programme production exceed those in broadcasting, we obtain negative vertical economies.

Our empirical results show that, although economies of scale exist both for programme production and for transmission, little evidence supports the existence of vertical economies nor encourages further integration between these two functions. (Our estimated parameters are listed in Appendix D.)

Reducing the burden is inevitable in achieving full-scale digitization of terrestrial broadcasting. A way to reduce the cost burden from investment in digital broadcasting is to share the transmission network with broadcasting stations. This is termed 'Network Sharing' and is repeatedly discussed in terms of investment for new telecommunications facilities (For example, Bjorkdahl and Bohlin, 2001). From the point of cost structure, separation is possible, and if network sharing is introduced, a considerable cost reduction can be realized. According to the estimation conducted in Mitomo and Ueda (2003), the initial investment can be reduced by 32 per cent if a transmission facility is shared by four stations. In an age of digital broadcasting, a new framework for programme distribution will facilitate the diffusion throughout the country. Network sharing can be introduced if the stations do not show strong economic effects from vertical integration.

In terms of cost structure, there appears to be no disadvantage in the separation of programme production and broadcasting. Such separation would be acceptable also as a first step towards the convergence of broadcasting and telecommunications.

CONCLUSION

Digital broadcasting is expected to realize a ubiquitous ICT environment by providing a variety of digital video, audio, text and other services. This chapter takes some first steps towards examining its economic effects with regard to scale, scope and vertical integration. With the approach of digital broadcasting, a burden of massive investment will be placed on local broadcasting stations. Obviously, due to a limited capability of investment, small or medium-sized broadcasting stations will have difficulty bearing the burden.

Reorganization could relieve the burden based on economic considerations. It would enlarge the operation of a broadcasting station or bring about a more efficient operation. It is difficult, however, to realize any actual reorganization because it would have to be accompanied by a drastic change in the industry. We should note that this study focuses merely on economic considerations; and of course, no industrial reorganization could be performed solely from an efficiency-based perspective.

We have tried to show evidence supporting future reorganization. The existence of economies of scale could be a basis for enlarging broadcasting areas in line with relaxation of the principle of abatement of mass media concentration. The results revealed a considerable level of scale economies in Japanese local broadcasting stations. Extending a broadcasting area has certain advantages in realizing a ubiquitous mobile ICT environment because it requires less frequent handovers of radio frequencies.

While horizontal integration is not conspicuous in the Japanese broadcasting industry, due to another aspect of the principle, vertical integration between programme production and transmission has so far characterized the industry. We assumed existence of economies of vertical integration, but it was not true concerning the cost structure of local stations. Little evidence could be found showing the synergetic economic effects of vertical integration existing in Japanese local broadcasting stations. If such economies do not exist, vertical separation is possible in an economic context. Furthermore, Japanese terrestrial broadcasting has long been vertically integrated, which has led to insufficient conditions for competition. With the approach of digital broadcasting, the telecommunication sector is expected to begin playing a complementary or substitute role in the distribution of digital programmes. Vertical separation between programme production and transmission could be a step towards future restructuring.

After separation, network sharing among the newly independent transmission divisions could be effective in reducing the cost burden of digitization and would act to restrain the competitive advantage these divisions might otherwise enjoy over telecommunications providers. In addition, the separate divisions would have access to an unrestricted choice of distribution media. Thus, vertical separation will pave the way for the convergence of broadcasting and telecommunications.

NOTES

1. According to the latest estimate by NAB(2003), the cost burden has increased to 6.4 billion yen per station.
2. However, key stations are allowed digital satellite broadcasting in addition to those media they have owned thus far.
3. The model and the results in this section are from Mitomo and Ueda (2003).

REFERENCES

Baumol W. J., J. C. Panzar and R. D. Willing (1988) *Contestable Markets and The Theory of Industry Structure*, New York: Harcourt Brace Jovanovich Inc.

Bjorkdahl J. and E. Bohlin (2001) 'Financial analysis of the Swedish 3G Licensees; where are the profits Version II', paper presented at the ITS 12th European regional conference.

Christensen, L. R., D. W. Jorgensen, and L. J. Lau (1973) 'Transcendental logarithmic production frontiers', *Review of Economics and Statistics*, **55**, pp. 28–45

Hayashi P. M., J. Y. Goo and W. C. Chamberlain (1997) 'Vertical economics; the case of US electric utility industry, 1983–87', *Southern Economic Journal*, **63**(3), 710–725.

Kaserman D. L. and J. W. Mayo (1991) 'The measurement of vertical economies and the efficient structure of the electric utility industry', *The Journal of Industrial Economics*, 483–502.

Kasuga N. and M. Shishikura (2004) 'Market structure and profit in the Japanese broadcasting industry', *Journal of Public Utility Economics*, **55**(3), 19–31 (in Japanese).

Mitomo H. (2000) 'Is broadcasting as interactive as telecommunications?: implications for digital BS broadcasting in Japan', European-Japanese Workshop in Bonn, Convergence of Telecommunications and Broadcasting: Policy Issues, University of Bonn.

Mitomo H. and Y. Ueda (2001) 'Vertical disintegration and network sharing in broadcasting digitization', paper presented at the Invisible College Workshop on the Convergence of Broadcasting and Telecommunication, Tokyo.

Mitomo, H. and Y. Ueda (2003) 'Vertical separation between programme production and transmission: network sharing in the Japanese broadcasting industry', *Communications & Strategies*, **52**, 239–255.

The National Association of Commercial Broadcasters in Japan (NAB) (1998) 'Business strategy by Japanese private broadcasters toward digital era in 2010' (in Japanese).

The National Association of Commercial Broadcasters in Japan (NAB) (2003) 'On the cost estimate of equipment for digitization of terrestrial commercial broadcasting' (in Japanese).

TV Research Weekly, 15 November 1999, 20 November 2000, and 5 November 2001 issues, Hoso Journal Co., Ltd. (in Japanese).

Ueda Y. and H. Mitomo (2002) 'Empirical analysis of integration and disintegration in the broadcasting industry', *JSICR (The Japan Society of Information and Communication Research) Annual Report*, pp. 13–25 (in Japanese).

Ueda Y. and H. Mitomo (2003a) 'Administrative services utilizing terrestrial digital broadcasting', *The Japan Association of Social Informatics*, **15**(2), 53–64 (in Japanese).

Ueda Y. and H. Mitomo (2003b) 'Empirical analysis of vertical separation and network sharing in the broadcasting industry', *Studies in Regional Science*, **33**(3), 159–172 (in Japanese).

Ueda Y. and H. Mitomo (2003c) 'Empirical analysis of spatial expansion in broadcasting industry', Proceedings of the 40th Annual Meeting of the Japan Section of the Regional Science Association, 509–516 (in Japanese).

Ueda Y., H. Takahashi and H. Mitomo (2004) 'Empirical analysis of economies of scale in broadcasting industry', *Journal of Information & Communication Research*, 72–73, 46–52 (in Japanese).

APPENDIX A: SPECIFICATION OF THE TRANSLOG COST FUNCTION

To estimate scale economies and the economies of vertical integration, we utilize the translog cost function which was first developed by Christensen *et al.* (1973). The cost function of broadcasting stations is formulated as follows:

$$\ln C = \alpha_0 + \sum_i \beta_i \ln W_i + \frac{1}{2}\sum_i \sum_j \beta_{ij} \ln W_i \ln W_j$$

$$+ \alpha_y \ln y + \frac{1}{2}\alpha_{yy}(\ln y)^2 + \sum_i \delta_{yi} \ln W_i \ln y$$

$$(i, j = L, K, O) \tag{13.A1}$$

where C: Total cost of broadcasting

y: Output of broadcasting
W_L: Price of labour input
W_K: Price of capital input
W_O: Price of other inputs
$\alpha_0, \beta_i, \beta_{ij}, \alpha_y, \alpha_{yy}$ and δ_{yi}:   Parameters

Although it is possible to estimate these equations directly, unfavourable phenomena such as multicollinearity may result in unstable estimates. Applying Shephard's lemma, we derive cost share equations with respect to input prices, and estimate them simultaneously with the cost function:

S_L, S_K, S_O are the shares of labour, capital and commodity costs in the total cost.

$$\frac{\partial \ln C}{\partial \ln W_i} = \frac{\partial C}{\partial W_i}\frac{W_i}{C} = \frac{W_i X_i}{C} = \alpha_i + \sum_i \beta_{ij} \ln W_i + \delta_{yi} \ln y \tag{13.A2}$$

$$\text{where } \sum_{i=1}^n S_i = 1 \quad \left(S_i \equiv \frac{W_i X_i}{C}\right) \tag{13.A3}$$

$$S_L = \beta_L + \beta_{LL}\ln W_L + \beta_{KL}\ln W_K + \beta_{OL}\ln W_O + \delta_{yL}\ln y + u_L$$
$$S_K = \beta_K + \beta_{LK}\ln W_L + \beta_{KK}\ln W_K + \beta_{OK}\ln W_O + \delta_{yK}\ln y + u_K$$
$$S_O = \beta_O + \beta_{LO}\ln W_L + \beta_{KO}\ln W_K + \beta_{OO}\ln W_O + \delta_{yO}\ln y + u_O \tag{13.A4}$$

Homogeneity conditions with respect to the factor input prices are:

$$\sum_i \alpha_i = 1, \quad \sum_i \beta_{ij} = \sum_j \beta_{ji} = \sum_i \delta_{yi} = 0 \qquad (13.A5)$$

Due to the conditions, only two share equations of the three are independent. The seemingly uncorrelated regression (SUR) method is applied to the estimation.

The Allen-Uzawa cross elasticity of substitution is given by:

$$\sigma_{ij} = \frac{\beta_{ij} + S_i S_j}{S_i S_j} \qquad i \neq j \qquad (13.A6)$$

The index of the scale economies can be represented as the increase in production cost due to the marginal increase in production:

$$SE \text{ (Scale Economies)} = \frac{\partial \ln C}{\partial \ln y} \qquad (13.A7)$$

Scale economies exist in the neighbourhood of the average of samples if *SE* is less than unity.

APPENDIX B: THE DEFINITION OF THE ECONOMIES OF VERTICAL INTEGRATION

Broadcasting involves two activities: programme production (an 'upstream' activity) and programme transmission (a 'downstream' activity). To investigate the separability of these activities, translog cost functions for 'Broadcasting' and 'Programme Production' are specified as follows:

Cost Function of Broadcasting

$$\ln C = \alpha_0 + \sum_i \alpha_i \ln W_i + \frac{1}{2} \sum_i \sum_j \beta_{ij} \ln W_i \ln W_j$$

$$+ \alpha_y \ln Y + \frac{1}{2} \beta_{YY} (\ln Y)^2 \qquad (i, j = l, k \text{ and } o) \qquad (13.B1)$$

where

C: Total cost of broadcasting
C_p: Cost of programme production
X: Output of programme production

Y: Output of broadcasting
W_p: Price of the input from programme production
W_l: Price of labour input
W_k: Price of capital input
W_o: Price of other inputs.

Applying Shepherd's lemma, we derive a cost-sharing equation with respect to the factor input prices and estimate them simultaneously.

$$S_l = \alpha_l + \beta_{ll}\ln W_l + \beta_{lk}\ln W_k + \beta_{lp}\ln W_p + \beta_{lo}\ln W_o + u_l$$
$$S_k = \alpha_k + \beta_{kl}\ln W_l + \beta_{kk}\ln W_k + \beta_{kp}\ln W_p + \beta_{ko}\ln W_o + u_k$$
$$S_p = \alpha_p + \beta_{pl}\ln W_l + \beta_{pk}\ln W_k + \beta_{pp}\ln W_p + \beta_{po}\ln W_o + u_p$$
$$S_o = \alpha_o + \beta_{ol}\ln W_l + \beta_{ok}\ln W_k + \beta_{op}\ln W_p + \beta_{oo}\ln W_o + u_o \quad (13.\text{B}3)$$

S_l, S_k, S_p, and S_o, denote the cost shares of labour, capital, programme production, and other inputs, respectively. Homogeneity conditions with respect to input prices are:

$$\alpha_l + \alpha_k + \alpha_p + \alpha_o = 1$$
$$\beta_{ll} + \beta_{kl} + \beta_{pl} + \beta_{ol} = 0$$
$$\beta_{lk} + \beta_{kk} + \beta_{pk} + \beta_{ok} = 0$$
$$\beta_{lp} + \beta_{kp} + \beta_{pp} + \beta_{op} = 0$$
$$\beta_{lo} + \beta_{ko} + \beta_{po} + \beta_{oo} = 0 \quad (13.\text{B}4)$$

Scale economies of programme production and broadcasting are given by the coefficients of the first-order terms of X and Y, respectively.

The test of vertical economies depends crucially on the separability assumption. To measure the degree of vertical integration, we have to calculate the following measure of vertical economy (VE):

$$VE \equiv 1 - [(\partial\ln C/\partial\ln Y)/(\partial\ln C_p/\partial\ln X)] \quad (13.\text{B}5)$$

where $\partial\ln C/\partial\ln Y$ and $\partial\ln C_p/\partial\ln X$ can be calculated from the estimated cost functions of broadcasting and programme production. If vertical integration does not reduce the total costs of broadcasting, VE should be negative (Hayashi *et al.*, 1997).

APPENDIX C: THE EMPIRICAL RESULTS: ECONOMIES OF SCALE

Table 13.A1   The estimated parameters

SE (scale economies)	The number of samples	Estimated coefficient		Standard deviation	t-value	p-value
0.4884	37	α_0	9.18505	0.09867	93.08020	[0.000]
		α_y	0.48840	0.06591	7.41006	[0.000]
		β_L	0.27651	0.02421	11.42230	[0.000]
		β_K	0.06235	0.00580	10.74820	[0.000]
		β_O	0.66114	0.02473	26.74000	[0.000]
		α_{yy}	0.11985	0.08551	1.40170	[0.161]
		δ_{yL}	-0.02963	0.02004	-1.47863	[0.139]
		δ_{yK}	-0.00518	0.00484	-1.07132	[0.284]
		δ_{yO}	0.03482	0.02056	1.69357	[0.090]
		β_{LL}	0.15575	0.05937	2.62325	[0.009]
		β_{LK}	0.01137	0.01399	0.81317	[0.416]
		β_{LO}	-0.16713	0.06040	-2.76689	[0.006]
		β_{KK}	0.02269	0.01182	1.92026	[0.055]
		β_{KO}	-0.03406	0.01815	-1.87644	[0.061]
		β_{OO}	0.20119	0.06555	3.06910	[0.002]

Table 13.A2   The Allen-Uzawa cross elasticity of substitution

Cross term	Estimated value	Standard deviation	t-value	p-value
$W_L * W_L$	-0.579378	0.78531	-0.73777	[0.461]
$W_K * W_K$	-9.20192	3.08869	-2.97923	[0.003]
$W_O * W_O$	-0.05227	0.15542	-0.33629	[0.737]
$W_L * W_K$	1.65964	0.82503	2.01162	[0.044]
$W_L * W_O$	0.08579	0.34200	0.25084	[0.802]
$W_K * W_O$	2.17183	0.46019	4.71947	[0.000]

APPENDIX D: THE EMPIRICAL RESULTS: ECONOMIES OF VERTICAL INTEGRATION

The estimated parameters

Table 13.A3   Cost function for broadcasting $R^2 = 0.9432$

	Parameter	Standard error	t-statistic	p-value
α_0	9.02362	0.029401	306.919	[0.000]
α_y	0.80160	0.02283	35.11180	[0.000]
α_l	0.21571	0.01839	11.72910	[0.000]
α_k	0.08469	0.00755	11.21410	[0.000]
α_p	0.29456	0.01247	23.61350	[0.000]
α_o	0.40504	0.01992	20.32920	[0.000]
β_{yy}	0.195271	0.05119	3.81465	[0.000]
β_{ll}	0.03198	0.02175	1.47032	[0.141]
β_{lk}	-0.00688	0.00813	-0.84585	[0.398]
β_{lp}	-0.04677	0.01276	-3.66631	[0.000]
β_{lo}	0.02167	0.02328	0.93054	[0.352]
β_{kk}	$2.35\mathrm{E}-03$	$6.21\mathrm{E}-03$	0.379077	[0.705]
β_{kp}	0.014364	$7.13\mathrm{E}-03$	2.01327	[0.044]
β_{ko}	$-9.84\mathrm{E}-03$	0.010754	-0.915252	[0.360]
β_{pp}	0.177384	0.01508	11.7626	[0.000]
β_{po}	-0.144976	0.016624	-8.72094	[0.000]
β_{oo}	0.133152	0.032761	4.06438	[0.000]

Table 13.A4 Cost function for programme production $R^2 = 0.7268$

	Parameter	Standard error	t-statistic	p-value
α_0	7.81486	0.06872	113.72	[0.000]
α_x	0.74976	0.03985	18.81550	[0.000]
α_l	0.66595	0.00927	71.82280	[0.000]
α_k	0.08120	0.00968	8.38780	[0.000]
α_o	0.25285	0.01083	23.34180	[0.000]
β_{xx}	0.12034	0.06925	1.73788	[0.082]
β_{ll}	−0.08297	0.02001	−4.14660	[0.000]
β_{lk}	−0.03254	0.00905	−3.59559	[0.000]
β_{lo}	0.11551	0.02246	5.14397	[0.000]
β_{kk}	0.00123	0.01379	0.08897	[0.929]
β_{ko}	0.03131	0.01171	2.67324	[0.008]
β_{oo}	−0.14683	0.02707	−5.42336	[0.000]

Vertical economies: $VE \equiv 1 - [(\partial\ln C/\partial\ln Y)/(\partial\ln C_p/\partial\ln X)] = -0.0691$
Scale economies: Broadcasting $\alpha_y = 0.80160$
Programme production $\alpha_x = 0.74976$

14. Comparative analysis of the market structure of broadcasting and telecommunications in Japan

Sumiko Asai

INTRODUCTION

In Japan, the Ministry of Internal Affairs and Communications (MIC) is in charge of setting regulations and policy for both the broadcasting and telecommunications industries. Telecommunications services are defined as the transmission of information to specific persons using electrical facilities. On the other hand, broadcasting services are defined as the transmission of information to non-specific persons through a wire or wireless network. Transmitting information through electrical networks is common to both telecommunications and broadcasting. However, differences in the market structure and market performance are observed between these industries.

In April 1985, Nippon Telegraph and Telephone Public Corporation (NTT Public Corp.) was privatized, and the entry of new players into the telecommunications market was authorized.[1] The prices of telecommunications services have decreased since the introduction of this competition mechanism, and users have had a lot of benefits. Although the telecommunications market is actually dominated by several large-scale carriers and may not be regarded as perfect competition, even large-scale carriers are unable to maintain high profit margins at present, due to price competition and the threat of potential entrants. In this respect, we may say that the Japanese telecommunications market has changed from a monopolistic to a competitive market.

Broadcasting consists of terrestrial broadcasting, satellite broadcasting and cable television (CATV). While satellite broadcasting and CATV have developed recently, the most familiar medium in Japan has been terrestrial broadcast services. Entry into the terrestrial broadcasting market has been regulated due to the scarcity of available radio frequencies, and no terrestrial broadcaster has yet exited from the market. In addition, its average ratio of profit to revenue has been at about 10 per cent and maintained at a high level,[2] although advertising, which is the primary revenue source for commercial terrestrial broadcasters, has recently remained static.[3] We may say that terrestrial broadcasters have not been fully conscious of competition under stable conditions,[4] as compared with telecommunications carriers. The first purpose of this chapter is to outline the market structure and consider the reasons why differences exist in market performance between the broadcasting and telecommunications industries.

Analogue networks were replaced by digital ones in the Japanese telecommunications market. Digitalization has not only increased the productivity of telecommunications carriers, but has also changed their business models and the market structure, as will be mentioned in the second section. In contrast, digital terrestrial broadcasting began in the metropolitan areas in December 2003, and analogue television services will be completely phased out by 2011. The second purpose of this article is to consider the impact of digitalization on the broadcasting market.

This chapter will undertake a comparative analysis of the market structure with respect to broadcasting and telecommunications, and consider the impact of digitalization on the market. The rest of this article is organized as follows: The second section explains the structure of both industries and the third section describes the impact of digitalization on the broadcasting market. The final section offers some concluding remarks.

MARKET STRUCTURE

Approach[5]

This chapter avoids the use of industry-specific terms and explains the market structure of the two industries, utilizing the following concept. The production process for goods and services is generally divided into several components and the interface between those components. For example, telephone services are largely composed of customer premises equipment (CPE), telephone lines and switching facilities. These components are connected according to the *de jure* standards, that is, technical standards established by the MIC and the International Telecommunications Union (ITU).

Recently, modularity has been discussed in the field of business administration. A module is a component of goods or services that is designed independently. Baldwin and Clark (1997, 2000) proposed the concept of modularity and analysed the computer industry in the U.S. In Japan, Fujimoto *et al.* (2001) analysed the automobile and other industries, using the concept of modularity. While modularity has been mainly discussed in relation to manufacturing, it has not commonly been applied to broadcasting so far. Therefore, in this paper, the term 'component' is used as a synonym for 'module'.[6]

There are several methods of combining components and the interfaces between components. First, a product is composed of one module. This is an integrated system and production is completed within a single firm.[7] Second, the production process can be divided into more than one component and the interface between these components is specified within the firm or firms concerned. The interface is closed or proprietary information. Third, the production process consists of more than one component and an open interface which a third-party can access. In this case, firms are given the choice to 'make or buy'. That is, they may produce some components themselves and procure other units from the market in order to provide final goods or services. When an interface is fixed with open information, firms are not required to pay attention to other components and the interface. Therefore, the transaction cost for combining several components is reduced. In addition, since firms do not need to produce all components themselves, small-scale firms that do not own a large amount of assets may go into business.

An open interface includes *de jure* and *de facto* standards. Telephone networks are interconnected using the *de jure* standard, and technical standards for television broadcasting are also established by public institutes and industrial organizations, prior to providing broadcast services. On the other hand, the Internet is a global network that uses the *de facto* standard, TCP/IP (Transmission Control Protocol/Internet Protocol), and we may also observe several *de facto* standards in the computer industry.[8] Next, the structure of broadcasting and telecommunications industries is described, using the concept of component and interface.

Structure of Broadcasting Industry

Broadcasting comprises terrestrial broadcasting, satellite broadcasting and CATV. In addition, satellite broadcasting is divided into broadcasting satellite (BS) and communications satellite (CS). In Japan, at the end of March 2005, the number of BS subscribers was about four times larger than the number of CS subscribers.[9] Satellite broadcasting service via a BS has not been provided in the UK and US, and the operation and penetration of BS are characteristics of the Japanese broadcasting market.

There are several reasons why BS broadcasting has operated in Japan. First, NHK is obliged to provide broadcasting services nationwide according to the Broadcasting Law. Since a satellite covers the entire country, NHK has an incentive to launch a satellite in order to fulfil the obligation of universal service. Second, NHK can afford to develop and operate BS, since NHK's revenues have been guaranteed due to the system of receiving a fee prescribed by the Broadcasting Law. Third, before the revision of the Broadcasting Law in 1989, broadcasting services via a CS had not been permitted in Japan. Therefore, NHK itself could not help launching a BS to provide satellite broadcasting services. Since NHK started BS broadcasting prior to CS broadcast services by Sky Perfect TV, we may say that NHK has enjoyed the first mover advantage.

The number of channels is different among terrestrial broadcasting, BS, CS and CATV, and their respective reception facilities are incompatible. Strictly speaking, the reception facilities for broadcasting via a CS launched at 110 degrees of east longitude (110° E CS broadcasting) can be utilized for BS broadcasting. However, 110° E CS broadcasting began in 2002, and the facilities for BS broadcasting that were installed before the start of 110° E CS broadcasting are incompatible with 110° E CS broadcasting. Viewers who were interested in television services had already subscribed to BS or other CS broadcasting services. As a result, the 110° E CS broadcasting service had only 202 000 subscribers at the end of March 2005. That is to say, Japanese broadcast industry has consisted of several incompatible platforms, and viewers are locked into the platform that they subscribe to first.

Terrestrial broadcasting largely consists of two components: transmission facilities and programmes. These are designed independently, but both are essential for broadcast service. As the Broadcasting Law assumes that broadcasters produce programmes, terrestrial broadcasters are principally engaged in both production and transmission of programmes. In this respect, terrestrial broadcasting is a vertical integration of hardware and television programmes. However, we may observe vertical separation between hardware and content in the terrestrial broadcasting market in two respects. The first factor is network affiliate contracts. In addition to NHK, which operates as a public broadcaster, there are 127 commercial broadcasters that provide terrestrial television service in Japan. Commercial broadcasters transmit their programmes, along with commercial messages, within their service areas. The Ministry expected that public broadcaster NHK would broadcast nationally, and that commercial broadcasters would play the role of transmitting local programmes. Therefore, in addition to the prohibition on mergers, the service area of commercial broadcasters is principally limited to their local prefectures from the standpoint of localism.[10] However, as the value of advertising media increases proportionally with the size of the audience, commercial broadcasters have an incentive to transmit their programmes into other areas. On the other hand, local stations do not possess adequate operational resources to produce all the programmes required to fill their television schedules. Therefore, so-called networks, that is large-scale broadcasters enter into affiliation contracts with several local stations to provide programmes and commercial messages. As local stations receive programmes from networks under their affiliation contracts, the ratio of locally produced programmes to total transmitted programmes is about 10 per cent on average. Although the Ministry has restricted service areas of broadcasters and has prohibited mergers in order to promote localism, it has not taken into account the interface between hardware and content across broadcasters. As a result, programmes produced by large-scale broadcasters in Tokyo have been transmitted nationally, and the localism intended by the Ministry has not been accomplished. Although terrestrial broadcasting is principally a vertical integration, hardware and content are separated in the local stations. In addition, in practice, broadcasters are integrated horizontally through their contracts, but the interface between broadcasters is not open, as the content of contract is not public information that a third-party can access.

The second factor in the vertical separation occurs in the relationship between terrestrial broadcasters and programme production companies. Although the first factor was the problem of relationship between large-scale broadcasters and local stations, the second factor is the problem of interface between hardware and content within a single broadcaster. While entry into the broadcasting market is regulated by the Ministry, there are no regulations controlling entry into the content market. In addition, since large-scale investment is not generally required to produce programme content, a lot of production companies have entered the market and are now operating.[11]

In Japan, not only local stations but also large-scale broadcasters often purchase television programmes from programme production companies for economic reasons. According to the Copyright Law, programme production companies are generally entitled to own their copyrights unless special agreements on the transfer of copyrights have been reached. However, as the number of large-scale terrestrial broadcasters in Japan has been limited to five, they are in a dominant bargaining position relative to the programme production companies. Consequently, it is often pointed out that broadcasters exercise copyrights without explicit contracts and thus production companies lose the opportunity to get additional revenue from a secondary use of their programmes.[12] Since few television programmes have been used secondarily so far,[13] policymakers had not paid attention to the interface between broadcasters and programme production companies[14]. However, as the opportunities for the secondary use of programmes are expected to increase as a result of the development of digitalization and the Internet, inappropriate agreements on copyrights will disturb the development of the content industry. To sum up, the Broadcasting Law assumes that terrestrial broadcasting is integrated vertically, but in reality the structure is vertically separated, and the design process of the interface between these components is not transparent.

In contrast, explicit separation of hardware and content has been implemented in the direct broadcasting satellite market. The revised Broadcasting Law of 1989 specified two additional kinds of broadcasters: facilities-supplying and programme-supplying broadcasters. Facilities-supplying broadcasters operate their communications satellites and distribute programmes produced by programme-supplying broadcasters. Programme-supplying broadcasters are engaged in the production of programmes, and entrust the distribution of their programmes to facilities-supplying broadcasters. Since programme-supplying broadcasting does not require a large amount of assets, more than one hundred firms in Japan have entered the satellite programmes market and vertical separation has contributed to an increase in content.

Although satellite broadcasting services are divided into hardware and content under the revised law, another component, that is, the operation of a subscriber management service, also exists. So-called platform providers collect charges from subscribers as well as select content. Unless platform providers undertake this role, programme-supplying broadcasters themselves have to collect information charges from their viewers, and viewers have to make contracts with each programme-supplying broadcaster. This becomes a burden to both programme-supplying broadcasters and their viewers.

Platform providers have appeared for economic reasons, and the interface between the platform providers and content producers has been specified by their contracts. On the other hand, since the platform provider is in a dominant bargaining position relative to programme-supplying broadcasters, problems such as taking undue advantage of their position and lack of transparency in content selection have been pointed out. Although the separation of hardware and content has been specified for the satellite broadcasting market, policymakers have failed to give due regard to the interface between the two components.[15] This is in contrast to a debate about the bottleneck, including subscriber management services, in Europe.[16] We may say that the Japanese broadcasting industrial structure consists of more than one component and a closed interface, and creates a situation that may lead to anti-competitive behaviour.

Structure of the Telecommunications Industry

Interconnection which corresponds to the interface between networks has been one of the major topics in Japanese telecommunications policy.

However, policymakers were not concerned about unbundling and open networks before the introduction of competition.

Before April 1985, Japanese domestic telecommunications had been operated by the NTT Public Corp. In addition to the provision of telecommunications services, the NTT Public Corp. jointly designed telecommunications facilities with certain manufacturers and provided telephone terminals for users. Technically the networks had a closed information configuration and the facilities were not compatible with foreign ones. Although those manufacturers were independent of the NTT Public Corp., from the standpoint of their shareholders and organization, the manufacture of facilities and the provision of services were vertically integrated,[17] with a closed interface.

Since the introduction of competition, the Ministry of Posts and Telecommunications (the MPT and the present MIC) has established rules for interconnections and required that the privatized NTT disclose technical information and cost data in order to promote competition. Under the interconnection rules, NTT's networks have been unbundled gradually, and network information has been open to the public.

Another approach to open networks is vertical separation of the incumbent.[18] Although the Ministry asserted that the separation of NTT was essential in order to promote fair competition, NTT had bitterly opposed the separation since privatization. This issue has long been under discussion, but the parties concerned finally achieved a consensus on the reorganization of NTT in December 1996. According to the agreement, in July 1999 NTT was separated into one long-distance company and two local-communications companies under a new holding company in order to create a level playing field between NTT and other carriers. Japanese telecommunications policy led to not only the unbundling of networks but also to the reorganization of NTT.[19]

The source of NTT's revenues in the 1980s was principally its standard telephone service. However, since the 1990s, part of the telephone service has been replaced by the Internet. The Internet consists of several components such as terminals, access and backbone networks, and servers. Since the protocol of the Internet is the *de facto* standard TCP/IP, the Internet has been divided into several components and open interfaces from its inception. In addition to the communications protocol, those protocols for producing content distributed over the Internet such as HTML (Hyper Text Markup Language) are also the *de facto* standards and open information. The existence of open language for the production of content has led to an abundance of content, and this has attracted people to the Internet. Thus, we can observe a positive feedback mechanism in operation.[20]

To sum up, when the Japanese telecommunications services were provided by a monopolistic firm, the firm was integrated vertically, and the network interface was closed. However, since April 1985, integrated networks have been divided into several components and the interface has been opened under regulation. In addition, the public telecommunications network has been partly replaced by the Internet protocol network which consists of modules and an open interface. This recent movement towards open networks did not result from regulation, but from the impact of the computer industry and the Internet. Although the means of open networks are different, we may say that open interfaces have promoted competition in the telecommunications market.

As telecommunications networks have changed, there have been corresponding changes in the organization of telecommunications carriers. Since a telecommunications network consists of several components and an open interface, carriers with small-scale networks can provide services worldwide by connecting with other carriers' networks. However, since the late 1990s, we have witnessed the integration of telecommunications carriers through mergers. KDDI and Softbank Corp. are examples of integration and diversification.[21] Since carriers can concentrate their resources in a limited field as business domains under an open interface, thereby acquiring competitive advantage, large-scale mergers and diversification in the telecommunications market seem to be contradictory. However, it is possible to demonstrate that there are economic reasons for integration and diversification.

Since the combination of components and open interface enables newcomers to enter the market, price competition will be promoted. In addition, since services provided under *de jure* and *de facto* standards are homogeneous, users view the services as substitutes for those provided by other carriers. As a result, severe price competition tends to occur. On the other hand, integrated carriers can provide one-stop shopping and various kinds of tariffs, and discounting across a range of services gives users an incentive to purchase a bundle of services from them. We may say that the purpose of a merger under an open interface is to acquire a competitive advantage through the enclosure of users.[22] In addition, technology that has realized a convergence of traditional telecommunications and computers has promoted a trend towards integration and alliances.

Terrestrial broadcasters have maintained vertical integration with a closed interface. On the other hand, although several telecommunications carriers have been integrated recently, they merged as a reaction to price competition under an open interface. The telecommunications and broadcasting markets have different reasons for integration, and the differences in the design of interfaces bring about the differences in the market structure and strategy.

IMPACT OF DIGITALIZATION

As the telecommunications markets have changed due to the digitalization of networks and the development of the Internet, so is digitalization expected to have an impact on broadcasting. First, digitalization enables broadcasters to provide multiple-channels. In Japan, a digital CS broadcaster provides more than one hundred channels. In the case of terrestrial broadcasting, one analogue channel corresponds to three digital channels, due to the efficient use of frequencies, as long as broadcasters do not provide high definition television (HDTV).[23]

The increase in the number of channels is expected to change the relationship between broadcasters and production companies. As mentioned above, broadcasters often purchase programmes from a third party. Since it is more difficult to produce all programmes in-house to satisfy the requirements of their television channels due to an increase in the number of channels, the purchase of programmes from production companies will inevitably increase. This will reduce the dominance of large-scale broadcasters over production companies. If the acquisition of attractive content becomes an important part of the management strategy of a broadcaster, production companies will not only increase their revenue but also find themselves on an equal footing to negotiate the transfer of copyrights.

Second, a portion of the public telecommunications networks has been transferred to computer networks, and various kinds of information such as voice, data and picture have recently been transmitted over the Internet. We will be able to enjoy digital television programmes over the Internet or on third-generation cellular phones. This is an example of the conversion of telecommunications, computers and broadcasting.

In Japan, terrestrial broadcasters that have provided analogue services are licensed for digital television, and the entry of new players into the digital terrestrial television market has not been authorized.[24] In addition, the Japanese broadcasting market has been segmented into terrestrial broadcasting, CS and BS broadcasting and CATV, and these sectors are all in the process of digitalization. In other words, while telecommunications networks tend to be integrated and the market structure has changed, due to the development of digitalization and the Internet Protocol (IP), the market structure of broadcasting is not basically expected to change as a result of digitalization.[25] Asai (2005) calculated the efficiency and productivity of Japanese terrestrial broadcasters during the period from 1997 to 2002, and indicated that yardstick competition did not work in the analogue broadcasting market. Improvement in market performance through competitive mechanisms is not expected in the terrestrial digital broadcasting market, as long as the market structure remains unchanged. However, just as the telecommunications business has changed rapidly due to the penetration of the Internet, so terrestrial broadcasters will be confronted with the Internet as a competitor. There is the possibility of change in the broadcasting market due to such outside influences.

CONCLUDING REMARKS

Terrestrial broadcasters are vertically integrated by law and horizontally integrated through their network affiliate contracts. In addition, since facilities for BS, CS and CATV are incompatible, viewers are locked into the broadcaster they had subscribed to first. This is contrary to the telecommunications business, which consists of several components and an open interface. We may say that the differences in the structure of these industries cause their differences in market performance.

However, broadcasting has characteristics which are not duplicated in the telecommunications market. First, content regulation has been implemented in the broadcasting market,[26] due to the fact that broadcasting has a great impact on society. On the other hand, although the impact of the Internet is also large, content distributed over the Internet has basically not been regulated or monitored. As the transmission of television programmes over the Internet becomes more common, the distinction between television and the Internet will become less apparent. It may be time to review the structure of the broadcasting industry and sector-specific regulations.

Second, telecommunications carriers, with the exception of NTT East and NTT West Companies can decide business domains at their own discretion,[27] unless they utilize radio frequencies.[28] On the other hand, one reason for entry regulations in the broadcasting market is to maintain the diversity of media and localism. Japanese television broadcasters had not been permitted to own other local stations in order to maintain media diversity and localism.[29] In addition, they have been prohibited from providing both radio and newspaper services. Although both the diversity of media and localism have a large impact on the market structure of broadcasting,[30] we need to examine the effects of such regulations.[31]

Finally, digitalization of Japanese telecommunications networks was completed in 1997.[32] Terrestrial broadcasting, two types of satellite broadcasting and CATV are currently in the process of digitalization. Therefore, the impact of digitalization on broadcasting cannot yet be determined, and it remains a matter to be further considered.

NOTES

1. The number of telecommunications carriers is 13 090 as of March 2005, according to an investigation by the MIC.
2. For the management of Japanese broadcasters, see the *Japanese Commercial Broadcasting Yearbook* published by the National Association of Commercial Broadcasters in Japan. The average ratio of profit to revenue was 3.1 per cent for all Japanese industries in 2002, according to the Basic Survey of Japanese Business Structure and Activities investigated by the Ministry of Economy, Trade and Industry. For the detail data, see http://www.meti.go.jp/
3. See the investigation by Dentsu Inc. http://www.dentsu.co.jp/
4. However, investment in digitalization and the growth of competition in the media industry will have adverse effects on the operation of broadcasters, especially local stations.
5. This approach has been adopted by Asai (2006a). Asai (2006a) analysed the market structure of telecommunications, employing the concept of modularity and interface. This article applies the approach to the broadcasting market. For the analysis of telecommunications in detail, see Asai (2006a).
6. A component in this chapter is also synonymous with a layer in the computer industry. In addition, since dividing the market along modular lines has been called 'unbundling' in the telecommunications and other service sectors, 'modularity' is synonymous with unbundling.
7. However, the first case may be extremely rare.
8. For example, the architecture of IBM-PC and operating system (OS) developed by Microsoft Corp.
9. While the number of BS subscribers is 14.82 million, the number of CS subscribers is 3.82 million, according to the *White Paper on Information and Communications in Japan* in 2005.
10. In the Kanto, Kansai and Tokai areas, commercial broadcasters are allowed to operate across several prefectures, in consideration of the economic connections between these areas.
11. The number of production companies and their activities has not been ascertained precisely. There are more than 120 companies that belong to the Association of All Japan Programme Production Companies.
12. For the anti-competitive conduct by broadcasters, see the Fair Trade Commission (2003).
13. In case of terrestrial television broadcasting, the ratio of programmes reused by other media was 7.7 per cent in 2002, according to an investigation conducted by the Institute for Posts and Telecommunications Policy.
14. Recently, the MIC, the Ministry of Economy, Trade and Industry and the Fair Trade Commission established study groups and considered the problems affecting content distribution.
15. The Ministry finally recognized the problems and introduced a guideline relating to the platform business in April 2003. However, the guideline is not enforceable.
16. Regarding the bottleneck and regulation of satellite television in Europe, see Nolan (1997) and Cave (1997).
17. Telecommunications carriers may complete a change to their tariffs by modifying the software currently installed. However, previously NTT Public Corp. had to modify switches as hardware in order to change the tariff. This meant that hardware and services were bundled.
18. However, the incumbent cannot be physically divided into a number of parts in the same way as networks are unbundled. Therefore, several interconnection rules are needed, even when the incumbent is separated vertically.
19. However, it is controversial whether the reorganization of NTT in 1999 was an example of separation or integration, since the holding company has owned a long-distance company and local-communications companies.

20. For example, NTT DoCoMo started i-mode cellular phone service in 1999. The function of i-mode service can be divided into three principal components: voice transmission, Internet access service and content on web sites. For content, NTT DoCoMo relied on third-party information providers and adopted the *de facto* standard HTML for producing content. As a result, many information providers have entered the content market, and a lot of content has contributed to the increase in i-mode subscribers. The business model of i-mode corresponds to the third case, as mentioned above. For an empirical test of the positive feedback mechanism that NTT DoCoMo has enjoyed, see Tanaka (2002).

21. International telecommunications carrier KDD, long-distance carrier DDI and cellular phone provider IDO merged into KDDI in 2000. Softbank Corp. has established subsidiaries and merged with other companies to provide various kinds of services such as ADSL, standard and IP telephone services, and e-commerce.

22. There is a possibility that large-scale carriers will exercise their market power derived from demand-side factors under the open interface. However, we will not take up the future problems on competition policy in this article.

23. When digital terrestrial broadcasters transmit HDTV programmes, the channel is limited to one, since the transmission of HDTV needs more frequencies compared with ordinary programmes.

24. When BS broadcasting services were introduced, new entries were authorized.

25. The research council on IT-related regulatory reform established within the Cabinet proposed regulatory reform relating to the relationship between broadcasting and telecommunications in December 2001. However, the proposal was bitterly opposed by terrestrial broadcasters, and a debate on the regulatory structure has not developed. For the report by the council, see http://www.kantei.go.jp/. For an opinion on the proposal, see the statement by the chairman of the National Association of Commercial Broadcasters in Japan.http://www.nab.co.jp/

26. However, there is doubt whether broadcasters have been fully conscious of content regulation.

27. NTT East and NTT West Companies that own essential facilities are prohibited from providing long-distance services in order to avoid anti-competitive behaviour.

28. When carriers utilize radio frequencies in order to provide their services, they have to get licences in advance.

29. The Ministry decided to allow mergers of local stations that faced financial difficulties, by way of exception, in 2004. See the press release from the MIC dated 17 March 2004. http://www.soumu.go.jp/

30. Asai (2006b) presented the existence of scale economies of television and radio services and scope economies between the two services, estimating the symmetric generalized McFadden cost function. The results of estimation imply that Japanese policymakers have placed priority on localism and media diversity rather than economic efficiency.

31. Pritchard (2002) examined the effect of media ownership rules in the U.S. He indicated that common ownership of a newspaper and a television station in a community did not result in a predictable pattern of news coverage in the community owned outlets.

32. In Japan, local analogue switching was completely replaced by local digital switching in 1997.

REFERENCES

Asai, S. (2005) 'Efficiency and productivity in the Japanese broadcasting market,' *Keio Communication Review*, **27**, 89–98.

Asai, S. (2006a) 'Changes in the interface and industry structure,' in R. Taplin and M. Wakui (eds), *Japanese Telecommunications Market and Policy in Transition*, Chapter 1, London: Routledge.

Asai, S. (2006b) 'Scale economies and scope economies in the Japanese Broadcasting Market,' *Information Economics and Policy*, forthcoming.

Baldwin, C. Y. and K. B. Clark (1997) 'Managing in an age of modularity,' *Harvard Business Review*, September/October, 84–93.

Baldwin, C. Y. and K. B. Clark (2000) *Design Rules, The Power of Modularity*, Vol. 1, Cambridge, MA: The MIT Press.

Cave, M. (1997) 'Regulating digital television in a convergent world,' *Telecommunications Policy*, 21(7), 575–596.

Fair Trade Commission (2003) *A Report on Study of Digital Content and Competition Policy*, http://www.ftc.go.jp/.

Fujimoto, T., A. Takeishi and Y. Aoshima (eds) (2001) *Business Architecture: Strategic Design of Products, Organization and Process*, Tokyo: Yuhikaku Co. (in Japanese).

Funada, M. and Y. Hasebe (eds) (2001) *Current issues on Broadcasting Law*, Tokyo: Yuhikaku Co. (in Japanese).

The Ministry of Internal Affairs and Communications (2004) *White Paper on Information and Communications in Japan*. http://www.soumu.go.jp/ (accessed September 2004).

Nakamura, K. (1999) 'Japan's TV Broadcasting in a Digital Environment,' *Telecommunications Policy*, 23, 307–316.

The National Association of Commercial Broadcasters in Japan (ed.) (2004) *The Japanese Commercial Broadcasting Yearbook*, Tokyo: Coken Publishing Inc. (in Japanese).

Nolan, D. (1997) 'Bottlenecks in pay television,' *Telecommunications Policy*, 21(7), 597–610.

Pritchard, D. (2002) 'Viewpoint diversity in cross-owned newspapers and television stations: a study of news coverage of the 2000 Presidential campaign,' Federal Communications Commission, Media Bureau Staff Research Paper, http://www.fcc.gov/ (accessed September 2004).

Takeishi, A. (2003) *The Division of Labour and Competition*, Tokyo: Yuhikaku Co. (in Japanese).

Tanaka, T. (2002) 'Econometric evidence of network externalities of the mobile phone industry in Japan,' *Mita Journal of Economics*, 95(3), 119–132 (in Japanese).

Index